IMPACT OF HIGH NIGHT TEMPERATURE ON PLANT BIOLOGY

Toward Sustainable Plant Adaptation to Climate Change

IMPACT OF HIGH NIGHT TEMPERATURE ON PLANT BIOLOGY

Toward Sustainable Plant Adaptation to Climate Change

Rajesh Kumar Singhal, PhD
Raju Bheemanahalli, PhD
Saurabh Pandey, PhD
Pratibha M. D., PhD
Editors

AAP APPLE ACADEMIC PRESS

First edition published 2025

Apple Academic Press Inc.
1265 Goldenrod Circle, NE,
Palm Bay, FL 32905 USA

760 Laurentian Drive, Unit 19,
Burlington, ON L7N 0A4, CANADA

CRC Press
2385 NW Executive Center Drive,
Suite 320, Boca Raton FL 33431

4 Park Square, Milton Park,
Abingdon, Oxon, OX14 4RN UK

Library and Archives Canada Cataloguing in Publication

Title: Impact of high night temperature on plant biology : toward sustainable plant adaptation to climate change / Rajesh Kumar Singhal, PhD, Raju Bheemanahalli, PhD, Saurabh Pandey, PhD, Pratibha M.D., PhD, editors.
Names: Singhal, Rajesh Kumar, editor. | Bheemanahalli, Raju, editor. | Pandey, Saurabh (Lecturer in agriculture), editor. | Prathibha M. D., editor.
Description: First edition. | Includes bibliographical references and index.
Identifiers: Canadiana (print) 20240445791 | Canadiana (ebook) 20240445880 | ISBN 9781774917336 (hardcover) | ISBN 9781774917343 (softcover) | ISBN 9781003503415 (ebook)
Subjects: LCSH: Plants—Effect of heat on. | LCSH: Vegetation and climate.
Classification: LCC QK755.5 .I47 2025 | DDC 581.4/2—dc23

Library of Congress Cataloging-in-Publication Data

Names: Singhal, Rajesh Kumar, editor. | Bheemanahalli, Raju, editor. | Pandey, Saurabh (Lecturer in agriculture) editor. | Prathibha M. D., editor.

Title: Impact of high night temperature on plant biology : toward sustainable plant adaptation to climate change / Rajesh Kumar Singhal, Raju Bheemanahalli, Saurabh Pandey, Pratibha M. D..

Description: First edition. | Palm Bay, FL : Apple Academic Press, 2025. | Includes bibliographical references and index. | Summary: "The persistence of higher night-time temperatures (HNT) over long periods during the crop cycle can potentially impact a wide range of growth and developmental stages. The major processes drastically affected by night-time warming, such as photosynthesis, respiration, etc., ultimately cause yield and economic losses. Despite this, HNT has not been significantly researched regarding its impact on plants. This book, Impact of High Night Temperature on Plant Biology, provides a comprehensive understanding of the physiology of night-time warming, trait responses, advanced tools for phenotyping, and alleviation and mitigation strategies. This book discusses HNT by starting with an introduction, responsible factors, and its effect on plant growth and development and then goes on to cover the impact of HNT on plant metabolism and crucial plant processes with reproductive stage effects. The book also observes the effect of the plant growth regulator on the plant under HNT along with omics interventions to mitigate the stress effects of HNT. The book focuses on advancing technologies and approaches that involve breeding, omics, transgenics, and bioengineering of metabolic pathways in unique ways, which aim to help develop climate-resilient crops under HNT. Key features of the book: Details current and future challenges of climate changes on food security in conjunction with HNT Highlights the effect of night-time warming on physiology, growth, and development of plants at different crop growth stages Explores plant processes or phenotype responses to night-time warming Provides an understanding of the crosstalk of night-time warming with other factors This book is among the first books focusing on the holistic response of plants to night-time warming. The assembled research information in this book will prove an excellent resource for agriculturists, plant scientists, climatologists, and research scholars"-- Provided by publisher.

Identifiers: LCCN 2024035020 (print) | LCCN 2024035021 (ebook) | ISBN 9781774917336 (hardback) | ISBN 9781774917343 (paperback) | ISBN 9781003503415 (ebook)

Subjects: LCSH: Plants--Effect of heat on. | Vegetation and climate. | Plant physiology.

Classification: LCC QK755.5 .I57 2025 (print) | LCC QK755.5 (ebook) | DDC 581.4/2--dc23/eng/20240826

LC record available at https://lccn.loc.gov/2024035020

LC ebook record available at https://lccn.loc.gov/2024035021

ISBN: 978-1-77491-733-6 (hbk)
ISBN: 978-1-77491-734-3 (pbk)
ISBN: 978-1-00350-341-5 (ebk)

About the Editors

Rajesh Kumar Singhal, PhD
Scientist (Plant Physiology), Division of Crop Improvement,
ICAR–Indian Grassland and Fodder Research Institute,
Jhansi, India

Rajesh Kumar Singhal, PhD, is a Scientist in the Division of Crop Improvement, ICAR-Indian Grassland and Fodder Research Institute, Jhansi (UP), India. He has eight years of experience in different areas of plant physiology aspects, such as drought stress tolerance in rice and wheat crop, seed priming, artificial night light pollution, root physiology, multiple abiotic stresses, and metabolism related to different crops. Currently, he is working on fodder oat improvement, oat modeling for climate change, and abiotic stress in forage crops. He has published more than 40 research and review papers in highly reputed journals. He has published more than 25 chapters with Elsevier, Springer, and Taylor & Francis. In addition, he has published an MCQ book, *A Quick Approach to Plant Physiology, Biochemistry and Biotechnology* (1st edition 2018; 2nd edition 2020). Dr. Singhal has received NTS, JRF and SRF during his degree program, awarded by the Indian Council of Agriculture Research. His work has been cited more than 565 times (14 h-index and 21 i10 index).

Raju Bheemanahalli, PhD
Assistant Research Professor, Department of Plant and Soil Sciences,
Mississippi State University, MS, USA

Raju Bheemanahalli, PhD, is an Assistant Research Professor in the Department of Plant and Soil Sciences at Mississippi State University, USA. He leads multidisciplinary research to understand how crops respond to stressors (drought, heat, cold, nutrients, and others) at different growth stages. His work has resulted in improved basic and applied knowledge of monocot and dicot crops in response to stressors and management. His research provided researchers with stress-tolerant donor lines and phenotyping tools required to breed climate-ready crops. The uniqueness of the information generated from the research program assisted in sustaining food and fiber production under stressful environments across the globe. He has been actively serving on editorial boards and a reviewer of numerous journals international journals. He has published more than 55 peer-reviewed journal articles in multiple highly reputed international journals.

Saurabh Pandey, PhD
Assistant Professor, Department of Agriculture in
Guru Nanak Dev University, Amritsar, Punjab, India

Saurabh Pandey, PhD, is currently working as Assistant Professor in the Department of Agriculture at Guru Nanak Dev University, Amritsar, Punjab, India. He has eight years of experience in different areas of plant biotechnology aspects, such as plant stress biology, plant virus interactions, abiotic stresses, bio-fortification, molecular marker studies, and metabolism related to tomato and rice crop. He has published more than 20 papers in reputed journals. Currently, he is working on introduction and biotic stress management of unconventional crops, like banana and apple, in the Punjab region for screening and analysis of banana germplasms for banana bunchy top virus (BBTV) tolerance and establishing screening methods and biomarkers for microbiological resources toward improved crop phosphorous (P) uptake under highly P-deficient soils in the Punjab region.

Pratibha M. D., PhD
Scientist (Plant Physiology), Division of Basic Sciences,
ICAR–Indian Institute of Horticultural Research, Bengaluru,
Karnataka, India

Pratibha M. D., PhD, is currently working as a Scientist in the Division of Basic Sciences at the ICAR-Indian Institute of Horticultural Research, Bengaluru, Karnataka, India. She has 10 years of experience in different areas of plant physiology, plant pathology, nematology, and molecular breeding in rice and vegetables. Her areas of specialization include genomics, metabolomics, root physiology, carbon isotope discrimination, and epicuticle wax under multiple biotic and abiotic stress conditions. Currently, she is working on identifying molecular mechanisms conferring combinatorial stress tolerance in brinjal (eggplant). She has also worked at Monsanto Holding Pvt. Ltd. as vegetable backcross breeder, contributing to the development of more than 15 hybrids in tomato and hot pepper.

Contents

Contributors

Arshi Naaj Afsana
Directorate of Forensic Science and Laboratory, Ranchi, Jharkhand India

Anjal
Department of Plant Physiology, G.B Pant University of Agriculture and Technology, Pantnagar, Uttarakhand, India

Gurdeep Bains
Department of Plant Physiology, G.B Pant University of Agriculture and Technology, Pantnagar, Uttarakhand, India

J Bharghavi
Department of Crop Physiology, Agricultural College, Warangal, PJTSAU, India

Mayank Pratap Singh Bangari
Department of Crop Physiology, University of Agricultural Sciences, Bengaluru, Karnataka, India

T. K. Behera
ICAR-Directorate of Weed Research, Jabalpur, Madhya Pradesh, India

Subhash Chander
ICAR-National Bureau of Plant Genetic Resources, New Delhi, India

Jyoti Chauhan
Noida International University, Noida, Uttar Pradesh, India

Basant Kumar Dadarwal
Institute of Agricultural Sciences, Banaras Hindu University, Varanasi, Uttar Pradesh, India

Prajjal Dey
Faculty of Agriculture, Sri Sri University, Cuttack, Odisha, India

K.K. Dwivedi
ICAR- Indian Grassland and Forage Research Institute, Jhansi, Uttar Pradesh, India

Sudhir Kumar Guru
Department of Plant Physiology, G.B Pant University of Agriculture and Technology, Pantnagar, Uttarakhand, India

Jessica Rene Hansdah
St. Xavier's College, Ranchi, Jharkhand, India

S G Harsha
Division of Basic Sciences, Indian Institute of Horticultural Research, Hesaraghatta Lake Post, Bengaluru, India
Department of Plant Physiology, Indira Gandhi Krishi Vishwavidyalaya, Raipur, Chhattisgarh, India

Vilakar Kayitha
Keladi Shivappa Nayaka University of Agricultural and Horticultural Sciences, (KSNUAHS), Shimoga, Karnataka, India

Yogesh Khokha
ICAR-Indian Grassland and Forage Research Institute, Jhansi, Uttar Pradesh, India

B. Kiran Kumar
Department of Botany, Osmania University, Hyderabad, Telangana, India

Rajeev Kumar
ICAR-Indian Institute of Vegetable Research, Varanasi, Uttar Pradesh, India

Asha Kumari
ICAR-Indian Agricultural Research Institute, Hazaribagh, Jharkhand, India

Dalpat Lal
Department of Agriculture, Jagan Nath University, Jaipur, Rajasthan, India

Teikur Majaw
Department of Botany, Plant Physiology & Biochemistry, CBS&H, Dr. Rajendra Prasad Central Agricultural University, Pusa, Bihar, India

Ashish Marathe
ICAR-National Institute of Biotic Stress Management, Raipur, Chhattisgarh, India

S.K. Meena
ICAR-Indian Grassland and Forage Research Institute, Jhansi, Uttar Pradesh, India

Ambika Prasad Mishra
Faculty of Agriculture, Sri Sri University, Cuttack, Odisha, India

Sunidhi Mishra
Department of Horticulture, College of Agriculture, Indira Gandhi Agriculture University, Raipur, Chhattisgarh, India

Udit Nandan Mishra
Faculty of Agriculture, Sri Sri University, Cuttack, Odisha, India

M. Nagaraju
Department of Botany, Osmania University, Hyderabad, Telangana, India

Karthik Nanaiah
Department of crop Physiology, University of Agricultural Sciences, Bengaluru, Karnataka, India

Saurabh Pandey
Department of Agriculture, Guru Nanak Dev University, Amritsar, Punjab, India

Megha Panwar
Department of Plant Physiology, G.B Pant University of Agriculture and Technology, Pantnagar, Uttarakhand, India

Ratna Kumar Pasala
ICAR-Indian Institute of Oilseeds Research, Hyderabad, Telangana, India

Rachana K. Pawar
Department of Crop Physiology, University of Agricultural Sciences, Bengaluru, India

Jyostna Rani Pradhan
Department of Botany, Plant Physiology & Biochemistry, CBS&H, Dr. Rajendra Prasad Central Agricultural University, Pusa, Bihar, India

Swati Puranik
Global Change Research Institute of the Czech Academy of Sciences, Bělidla, Brno, Czechia

M Rajashekhar
Agricultural Entomology, PJTSAU, Hyderabad, Telangana, India

B. Rajashekar
KVK, Palem, PJTSAU, Hyderabad, Telangana, India

B Rakesh
Department of Soil Science and Agricultural Chemistry, TNAU, Tamil Nadu, India

Sudarshana Ranjan
Department of Plant Physiology, G.B Pant University of Agriculture and Technology, Pantnagar, Uttarakhand, India

Debanjana Saha
Department of Biotechnology, Centurion University of Technology and Management, Jatni, Bhubaneswar, India

Pranav Pankaj Sahu
Global Change Research Institute of the Czech Academy of Sciences, Bělidla, Brno, Czechia

Data Ram Saini
Institute of Agricultural Sciences, Banaras Hindu University, Varanasi, Uttar Pradesh, India

S Saranya
GD Goenka University, Gurugram, Haryana, India

Asha Sastya
Department of crop Physiology, University of Agricultural Sciences, Bengaluru, Karnataka, India

Eetela Sathyanarayana
Department of Soil Science and Agricultural Chemistry, Agricultural College, Palem, PJTSAU, India

Neha Sharma
Amity Institute of Biotechnology, Amity University, Noida, Uttar Pradesh, India

H S Sheela
Division of Basic Sciences, Indian Institute of Horticultural Research, Hesaraghatta Lake Post, Bengaluru, India

K S Shivashankara
Division of Basic Sciences, Indian Institute of Horticultural Research, Bengaluru, India

P. K. Sing
ICAR-Directorate of Weed Research, Jabalpur, Madhya Pradesh, India

Hemlata Singh
Department of Botany, Plant Physiology and Biochemistry, CBS&H, Dr. Rajendra Prasad Central Agricultural University, Pusa, Bihar, India

Rajesh Kumar Singhal
ICAR-Indian Grassland and Forage Research Institute, Jhansi, Uttar Pradesh, India

Sneha
Department of Plant Physiology, Kerala Agricultural University, Vellayani, Kerala, India

Shobha Sondhi
ICAR-Directorate of Weed Research, Jabalpur, Madhya Pradesh, India

Dasari Sreekanth
ICAR-Directorate of Weed Research, Jabalpur, Madhya Pradesh, India

Teja V Surya
Department of Soil Science and Agricultural Chemistry, Assam Agricultural University, Jorhat, India

K Charan Teja
Department Agronomy, Agricultural College, Palem, PJTSAU, India

Saideep Thallapall
Forest College and Research Institute, Mulugu, Telangana, India

Pawar Deepak Vishwanath
ICAR-Directorate of Weed Research, Jabalpur, Madhya Pradesh, India

Nitin Yadav
Institute of Agricultural Sciences, Banaras Hindu University, Varanasi, Uttar Pradesh, India

Abbreviations

ABA	abscisic acid
AGPase	ADP-glucose pyrophosphorylase
APX	ascorbate peroxidase
AREB	ABA-responsive element-binding proteins
BRs	brassinosteroids
CaM	calmodulin
CAM	Crassulacean acid metabolism
CAT	catalase
CBK3	CAM-binding protein kinase 3
CDPK	calcium-dependent proteins kinases
CE-MS	capillary electrophoresis tandem mass spectrometry
CFCs	chlorofluorocarbons
CKX	cytokinin oxidase/dehydrogenase
CRF	cytokinin response factor
CTK	cytokinins
CWI	cell wall invertase
DEGs	differentially expressed genes
DT	daytime temperature
ERFs	ethylene-responsive factors
ET	ethylene
ETR	excitation-to-emission ratio
FAD	fatty acid desaturase
G-1-P	glucose-1-phosphate
GA	gibberellic acids
GABA	ϒ-aminobutyric acid
GB	glycine betaine
GBSS	granule-bound starch synthase
GC-MS	gas chromatography tandem mass spectrometry
GDH	glutamate dehydrogenase
GHGs	greenhouse gases
GR	glutathione reductase
GS	glutamine synthetase
GWAS	genome-wide association studies
H_2O_2	hydrogen peroxide
HCFCs	hydrochloroflourocarbons
HK	histidine kinase

HNT	high night-time temperature
HPLC	high performance liquid chromatography
HPT	histidine-containing phospho-transmitter
HSFs	heat stress factors
HSPs	heat-shock proteins
HSR	heat-shock response
IAA	indole-3-acetic acid
IPCC	Intergovernmental Panel on Climate Change
IPT	isopentenyl transferases
IRRI	International Rice Research Institute
JA	jasmonic acid
LCB	long-chain bases
LC-MS	liquid chromatography tandem mass spectrometry
LNT	low night temperature
m	meter
MALDI	matrix-assisted laser desorption ionization
MDHAR	monodehydroascorbate reductase
mGWAS	metabolome-assisted genome-wide association study
miRNAs	microRNAs
MTS	membrane thermal stability
NGS	next generation sequencing
NMR	nuclear magnetic resonance
NPR	non-expresser of pathogenesis-related
NR	nitrate reductase
NSC	non-structural carbohydrates
NT	night-time temperature
NTAT	night-time air temperature
PA	phosphatidic acid
PC	phosphatidylcholine
PE	phosphatidyl ethanolamine
PG	phosphatidyl glycerol
PGRs	plant growth regulators
PI	phosphatidyl inositol
PIF	phytochrome-interacting factor
POD	peroxidase
PROX	proline oxidase
PS	phosphatidyl serine
QTL	quantitative trait loci
RILs	recombinant inbred lines
ROS	reactive oxygen species
RR	response regulator
RZT	root-zone temperature

SA	salicylic acid
SAM	shoot apical meristem
SBE	starch branching enzyme-I
SFC	super critical fluid chromatography
SMRT	single-molecule real-time
SNPs	single nucleotide polymorphisms
SOD	superoxide dismutase
SPP	sucrose phosphate phosphatase
SPS	sucrose phosphate synthase
SUS	sucrose synthase
TAG	triacylglycerol
TCA	tricarboxylic acid cycle
UNEP	United Nations Environment Program
UPLC	ultra performance liquid chromatography
UV	ultraviolet
VPD	vapor pressure deficit

Foreword

Climate change is one of the most significant challenges of our time, and its impact is felt across the globe. One of the most alarming consequences of climate change is the temperature rise, especially at night. As the Earth's temperature continues to rise, plants are among the most vulnerable organisms facing the consequences of these changes. High night-time temperatures can have a harmful effect on plant growth and development, which can, in turn, affect food security, biodiversity, and ecosystem stability.

I am Raju Bheemanahalli, and I have been working on understanding the plant responses to abiotic stresses at different developmental stages to choose a plant that will grow optimally under extreme environmental conditions. It is my great pleasure to introduce the book titled *Impact of High Night Temperature on Plant Biology: Toward Sustainable Plant Adaptation to Climate Change*. This book focuses on the impact of high night temperatures on plant biology and provides a deeper understanding of the natural and anthropogenic factors responsible for it. The book provides a comprehensive overview of the current state of knowledge on the effects of high night temperatures on plant growth and development, nutrient cycling, metabolism, and reproductive growth, as well as on yield and quality attributes.

This book comprises 11 chapters that delve into different aspects of how plants respond to elevated night-time temperatures. Chapter 1 provides an overview of the natural and anthropogenic factors responsible for night-time warming, including the role of greenhouse gases and land use change. Chapter 2 discusses how elevated CO_2 levels and night-time warming affect plant growth and development. Chapter 3 analyzes the impact of warmer nights on plant growth and development, while chapter 4 provides an overview of the responses of key plant processes and growth stages to night warming. Chapters 5 to 7 delve into the impact of high night-time temperatures on nutrient cycling, metabolism, respiration, and the lipidome. Chapters 8 and 9 go into the effect of night-time warming on the reproductive, yield, and quality attributes. The final chapter highlights the role of plant growth regulators and omics technology in combating the negative impacts of high night temperatures on plant biology.

The contributors to this book bring a wealth of expertise and knowledge to the subject matter. They highlight the need for sustainable adaptation strategies to mitigate the effects of climate change on plant biology. As we continue to witness changes in the global climate, this book provides valuable insights into the impact of high night temperatures on plant sciences and the measures that can be taken to improve plant adaptation and resilience. We hope that this book will be a valuable resource for students, researchers, and professionals in the fields of agriculture, plant biology, plant science, ecology, and environmental science. I hope that this book will inspire further research in this area and contribute to the development of sustainable solutions for climate change adaptation.

Our sincere thanks go to the authors for their excellent contributions and the publishers for bringing the book to fruition.

—Dr. Raju Bheemanahalli
Department of Plant and Soil Sciences,
Mississippi State University, Mississippi, USA

Preface

The United Nations Sustainable Development Goals (SDGs) set a pressing issue for agriculture researchers to remodel the global food, environmental, and nutrition security sector under future climate change and extreme environment scenarios. The global food demand is expected to rise on average by 70–85% as the population reaches over 9 billion in 2050. Climate-change-induced weather alterations and increased frequencies of natural disasters have worsened the scenario by impacting global crop productivity. Currently, agricultural sector and economy face enormous challenges, such as growing demand for food with reduced crop productivity from unpredictable rainfall, water deficit, high salinity, and temperature fluctuations.

The persistence of higher night-time temperature (HNT) over long periods during the crop cycle can potentially impact a wide range of growth and developmental stages. Farm to global scale has already experienced a faster rate of rising in minimum night-time than the daytime temperatures. This results in a narrow diurnal temperature amplitude. HNT casts a foreboding impact on global food and nutritional security as it affects major crops' genetic potential and genetic gain. Research on different field crops has concluded that the HNT impacts physiology, growth, reproductive, and seed development, ultimately associated with lower yield and quality. The major processes drastically affected by night-time warming are photosynthesis, respiration, biophysical properties of leaves, carbohydrate metabolism, source–sink relationship, and nutrition, which ultimately cause yield and economic losses. Asymmetric increase in the daytime and its impact on crop production has been primarily explored, but a similar emphasis has not been given to night-time warming. There is a need for compressive understanding physiology of night-time warming, traits responses, advanced tools for phenotyping, and alleviation strategies.

With this background, *Impact of High Night Temperature on Plant Biology: Toward Sustainable Plant Adaptation to Climate Change* compiles the current aspects of HNT in four dimensions. First, we have introduced HNT, responsible factors, and its effect on plant growth and development. Further, the HNT impact on plant metabolism and crucial plant processes was explored with reproductive stage effects. In the last part, we have discussed the plant growth regulator's effect on the plant under HNT with omics interventions to mitigate the stress effects of HNT. The text focuses on advancing technologies and approaches, such as breeding, omics, transgenic, and bioengineering of metabolic pathways in unique ways, which helps us develop climate-resilient crops under HNT.

We believe that this book is unique in content and, to our awareness, is the first book that focuses on plants' holistic response to night-time warming. Also, the assembled research information in this book is an excellent reference book for agriculturists, plant scientists, climatologists, and research scholars.

Introduction

Sustainable crop production to feed the increasing world population is a major global priority under rapidly changing environments. The persistence of higher night-time (HNT) temperatures over long periods during the crop cycle can potentially impact a wide range of growth and developmental stages. Crop production in farms on the global scale has already experienced a faster rate of rising minimum night-time than the daytime temperatures. This results in a narrow diurnal temperature amplitude. HNT casts a foreboding shadow on global food and nutritional security as it is affecting the genetic potential and genetic gain of major crops. Research on different field crops has so far concluded that the HNT impacts physiology, growth, reproductive, and seed development, ultimately associated with lower yield and quality attributes. The major processes that were drastically affected by night-time warming are photosynthesis, respiration, biophysical properties of leaves, carbohydrate metabolism, source-sink relationship, and nutrition, which ultimately causes huge yield penalties and economic losses. Asymmetric increase in the daytime and its impact on crop production has been largely explored but a similar emphasis has not been given to night-time warming.

There is a need for a comprehensive understanding of the physiology of night-time warming, traits responses, advanced tools for phenotyping, and alleviation strategies. We hope this book has covered all the different aspects of night warming, status, nature, responses on physiology, biochemical, and molecular levels. Also, the central aim of this book is to provide in-depth understanding of night warming, which will assist in reducing the yield gap in near future. Also, the discussion of advanced technologies, such as omics technologies, which help mitigate the plant stresses in cultivated crops, will introduce readers to state-of-the-art developments unique to this field of study.

PART I

NIGHT-TIME WARMING: AN INTRODUCTION

CHAPTER 1

Natural and Anthropogenic Factors Responsible for Night-time Warming

ARSHI NAAJ AFSANA[1], JESSICA RENE HANSDAH[2], and ASHA KUMARI[3]

[1]*Directorate of Forensic Science and Laboratory, Ranchi, Jharkhand, India*

[2]*St. Xavier's College, Ranchi, Jharkhand, India*

[3]*ICAR—Indian Agricultural Research Institute, Hazaribagh, Jharkhand, India*

ABSTRACT

Plant biology highly depends on the physicochemical conditions around it. One of the most concerning conditions prevailing nowadays is the severe impact of night-time warming on plant growth and development. Extensive studies done for many years have revealed the changes in day and night-time temperatures through these years. Human health is also in danger due to this shift in global temperature. There has been a change in the heat wave pattern of different regions as a result of elevated global temperature. This elevated temperature is the consequence of anthropogenic emissions and natural phenomena. The greater the night-time is warm, the wetter the climate is, and this phenomenon is responsible for variations in plant growth and their overall ecology including the members of the food chain. On the contrary, during daytime, the warming is accompanied by dry conditions, which increases heat stress (HS) in the species and causes them to dehydrate at a greater pace. The reason

Impact of High Night Temperature on Plant Biology: Toward Sustainable Plant Adaptation to Climate Change. Rajesh Kumar Singhal, Raju Bheemanahalli, Saurabh Pandey, and Pratibha M. D. (Eds.)

behind night-time warming is associated with increased cloud cover. The clouds retain the heat during the daytime and release them during the night. On the contrary, during the day, cloud cover is low which enables the heat to rapidly reach the surface but this is short-lived as the heat is lost as rapidly. This causes a change in precipitation and humidity, triggering physiological functions to fluctuate, to bring about changes in vegetation yield and leaf area index of the plants. Understanding the impact on plant physiology will help us toward sustainable plant adaptations to these climate changes (CCs).

1.1 INTRODUCTION

Worldwide temperature rise has rendered the whole living world in danger. The flora–fauna, ecosystems, food chains, climate conditions, ecological balance, and many more components of the globe have now become vulnerable to great damage. Not only animals but also plants experience the changes in temperatures. Natural phenomena and human activities are the principal causes of global warming. The reasons behind the global rise in annual temperature have been dealt with in this chapter. HS is a prime area of concern to agriculture that also reduces crop yield. We, as humans, can seek a shady place to hide from a sunny area but plants do not act so, rather they are evolved to fight HS differently. However, the mechanism of how they receive, transmit, and respond to HS remains as a puzzle.

Temperature is a principal physical parameter that controls plant growth and survival and thus it affects plant productivity. For instance, a huge decline in productivity has been predicted in the case of wheat (−6%) for every 1°C rise in global temperature.[1] The plants adapt to these changes by synthesizing protective enzymes or, proteins are known as chaperones or, heat shock proteins (HSPs) and accordingly they modulate growth responses. Understanding how plants act upon such stress is very important for developing crops that can tolerate rising average temperatures without hampering productivity. So, it is very crucial to work out the cascade of how plants perceive contemporaneous changes in temperature and then how they respond to triggered chemical pathways to safeguard themselves by producing protective HSPs.

1.2 CAUSES FOR NIGHT-TIME WARMING

With climate change (CC), the average temperature is anticipated to rise by 1.0–3.7°C by the turn of the century.[2] The night temperature will be more impacted by CC than the daytime temperature.[3] Since national temperature recordkeeping started in 1985, the global standard daily minimum temperature has raised by 0.8°C per century, equated to the increase in the daily maximum temperature by 0.4°C.[4] Night-time warming can be caused by various natural factors. Some of the factors are discussed here.

1.2.1 NATURAL FACTORS

a. **Global warming:** It traps the extra solar energy at any time which raised the levels of humidity. The moisture present in the humid air later forms clouds which trap the heat and reradiate during the night causing high night-time temperatures (HNTs).[5] Plants are hampered in two ways when we talk about global warming—firstly due to the expansion of greenhouse gases (GHGs) and secondly due to CC. This happens because of the increased level of air pollutants in the atmosphere, consequently reaching the atmospheric layer and imposing the greenhouse effect. To keep a check on warming, plants absorb some of these gases on Earth's surface so that they do not reach the atmosphere. Such changes in weather patterns also have great impacts on the balanced function of vegetation. This also affects precipitation and drought periods, hence changes in growth patterns, flowering time, and water use of plants. Due to hot temperatures, in an urge to get cooled down, plants grow taller with their stalks becoming longer, shrinking leaves, and growing far away from other plants. Because of the increased distance between them plants being unsupported become unstable and likely to bend and break.[6]

b. **Cloud cover:** Pollution, the natural circulation pattern of oceans, etc. cause cloudiness, consequently trapping the heat from the Earth back to space.[7] As stated by Boucher et al.,[8] clouds have a very complex relationship with temperature. Giving back short-wave radiation into space is negative feedback whereas positive feedback involves retaining long-wave radiation close to the Earth's surface.[9] To understand past and future CC, the feedback relationships between temperature and cloud cover must be dealt with as discussed in Ref.8.

c. **Short-wave Radiation:** The short-wave radiation that reaches the Earth's surface also adds to elevation of temperature. Short-wave radiations have a higher amount of energy in contrast to long-wave radiations. As the Sun has enormous energy, it gives us short-wave radiations in the form of ultraviolet (UV) rays that add up to the temperature rise.[10] As mentioned above it is significantly trapped by the cloud cover also.

d. **Greenhouse gases (GHGs):** Natural processes maintain equilibrium between the incoming and outgoing solar energy, thus, in turn, maintaining optimum temperature but certain GHGs such as CO_2, NOx, CH_4, etc., in the Earth's atmosphere, and thus retaining the heat that they absorbed.[11] GHGs are warming the nights more than days. This seems to continue unless the emissions are not managed at a priority.

e. **Forests:** Even though forests have many benefits but they also contribute to global warming by releasing water vapor during photosynthesis. Studies have proved that water vapor absorbs more heat than CO_2.

f. **Forest fires:** Forest fires are increasing with time due to dry weather which is rapidly increasing the earth's temperature. The forest fire leads to the combustion of CO_2 stored in the trees. This adds to the atmospheric GHGs.

1.2.2 *ANTHROPOGENIC FACTORS*

a. **Greenhouse gases:** Not only naturally but GHGs are emitted due to human ventures and contributed higher amount of GHGs. Chlorofluorocarbons (CFCs) have been released in large quantities. They are used as propellants in aerosol cans, in the manufacture of plastic foams for cushions and other products, in the cooling coils of refrigerators and air conditioners, as fire extinguishing materials, and as solvents for cleaning (United Nations Environment Programme). Some compounds such as hydrochlorofluorocarbons (HCFCs) are being introduced as CFC substitutes. It causes less damage to the ozone layer but is still a threat as a GHG.

b. **Electric heaters:** The use of Electric heaters has been proven effective in warming. The temperature rise may be differently responsive to the number of heaters being used and the angle of placement of each heater. It also depends on the type of heater, for example, a convection heater is used to warm the air. The warm air tends to move upward which in turn circulates the heat in a closed environment. The other type of heater is the infrared heater which warms the objects around it and heats the environment rapidly in contrast to convection heaters.

c. **Wind machines:** These are usually used in vineyards to provide a warm circumstance during night-time. These are large fans that pull the hot air and provide it in the direction that we want. These are dependent upon electricity to power them.

d. **Fire pots:** Pots lit by fuels are one of the most ancient ways to prevent plants from freezing in the orchard. They require less maintenance and are cheaper than heaters and wind machines but are not environmentally friendly as combustion causes air pollution.

e. **Hand warmers:** Most commercially available hand warmers are created in such a way that they can be utilized just once. They generate heat using exothermic reaction which involves iron being oxidized or burning butane. This method is cheaper but less effective than others.

f. **Reflective curtains:** These curtains act like artificial clouds that absorb the heat and reemit it back to the field. These are effective but may hamper many natural processes as they act as a physical barrier to organisms that help the ecology run smoothly.

g. **Radiant objects:** These are any objects that store solar energy in the daytime and radiate it at night. This method requires no electricity but is dependent on many climatic factors and thus not very efficient.[12]

1.3 MOLECULAR MECHANISM OF THERMOTOLERANCE

As the day-night temperatures fluctuate, the response of plants toward this also fluctuates.[13–15] The HS experienced by the plant during noon is much more likely to support survival whereas it is the opposite during the night-time. Studies on diurnal variation in temperature in *Arabidopsis thaliana* showed that plants show diurnal variation in heat resistance or, better to be called thermotolerance. This protects the plant from the warmest phases of the day. It also saves energy that is invested to produce HSPs during comparatively cooler nights.

1.3.1 LIGHT-MEDIATED HEAT STRESS RESPONSE

It has also been proven that light exposure set off heat resistance in plants. This heat resistance property is lost in darkness and is only regained when exposed to light again. Scientists have discovered that several genes involved in chloroplast formation also affect the response of plants to HS. Hence, it was noted that chloroplast genes were co-related to the HS response and showed a better response during the daytime. The chloroplast communicates a signal in responding to light and triggers expression of genes in the nucleus to induce tolerance ability against HS. The photosynthetic electron transport chain (ETC) is thought to be the initiator of the signaling process. This signal is then passed on to the nucleus for the activation of other transcription factors that directly or indirectly switch on other genes involved in HS response by synthesizing HSPs. However, the exact signaling molecule, which is initiating the response through the photosynthetic ETC is not known yet. It could be H_2O_2 which regulates initiation of signaling and moves toward the nucleus from the chloroplast.

1.3.2 HEAT SHOCK PROTEINS FOR THERMOTOLERANCE IN PLANTS

Lots of studies in eukaryotes have revealed that for HS response there are conserved heat shock transcription factors (HSFs) that majorly regulate the genes encoding molecular chaperones and other stress proteins.[16]

Higher plant genomes contain a large HSF family in comparison to other eukaryotes.[17,18] According to Nover et al.[17] and Wang et al.[19] in Arabidopsis and rice at least 21 and 25 HSF homologues are expressed. Whereas the genomes of invertebrates and vertebrates contain one and four HSFs respectively.[20] The large HSF family in higher plants indicates its relationship with the tolerance of sessile organisms that face diverse environmental challenges including extreme temperature, high light intensity, drought, flood, and soil salinity. According to Hayhoe et al.[21] it is very important to understand the role and complexity of the HSF network in higher plants in the context of global CC, which may lead to a significant increase in extreme heat frequency and duration in some regions. The signaling molecules generated during stress may trigger the core transcription factors and then induce an appropriate transcriptional response. It has been shown that in different biotic and abiotic stress conditions, H_2O_2 serves as a signaling molecule.[22] Nover et al.[17] identified putative HSFs in *Arabidopsis* with the aid of whole genome sequence data of the plant; they are categorized into three major classes (class A, B, and C) and 14 groups (A1–9, B1–4, and C1). *Arabidopsis* has four homologs of class A1 HSFA1 genes and is supposed to regulate heat shock response (HSR). In quadruple knockout (QK) and four triple knockout mutants of the plant, an extensive study was carried out to detect their functions. The outcomes show the relationships between HSFA1s and other previously identified components that contribute to thermotolerance. Consequently, HSFA1 group plays a critical role in HSR and they also support growth and development. Numerous modifications have been noted in morphology, and are usually associated with retardation of growth in the quadruple. However, no such alterations were observed in the triple KO mutants. Through this model, it has been analyzed that the activation of HSFA1s takes place upon heat shock treatment due to an increase in signaling molecules such asH_2O_2 and calcium,[23,24] eventually passing on the signal through molecules such as calmodulin (CaM) and CaM-binding protein kinase 3 (CBK3).[25] More than 200HSU genes are activated HSFA1s up-regulate the transcription including HSFA2, HSFA7a/7b andDREB2A. Under heat stress (HS) conditions DREB2A controls the expression of HSFA3.[26,27] Based on the annotation in MAPMAN,[28] the HS-upregulated genes are classified into functional categories as shown in Table 1.1. However, when to activate the genes that produce HSPs remains a mystery.

TABLE 1.1 Classification of HSFA1-Dependent HSU Genes.

S. No.	Function classification	Gene name (AGI#)
1.	HS	HSP15.7-CI (AT5G37670), HSP17.6A-CI (AT1G59860), HSP17.6C-CI(AT1G53540),HSP17.6B-CI (AT2G29500), HSP17.8-CI (AT1G07400), HSP17.6-CII (AT5G12020),
		HSP17.4-CIII (AT1G54050), HSP17.4 (AT3G46230), HSP17.6A (AT5G12030), HSP21(AT4G27670), HSP22 (AT4G10250), mtHSP23.6 (AT4G25200), HSP26.5-P (AT1G 52560),HSP60(AT3G23990),HSP70(AT3G12580), HSC70-2 (AT5G02490), HSP70B (AT1G16030),
		HSP70T-2 (AT2G32120), cpHSC70-1 (AT4G24280), mtHSC70-2 (AT5G09590), HSP90.1(AT5G52640), HSP90.2 (AT5G56030), HSC70-3 (AT3G09440), HSP101 (AT1G74310), CLPB-M(AT2G25140), ROF1 (AT3G25230), ROF2 (AT5G48570), FKBP-TYPE PPIASE (AT4G25340),HSA32 (AT4G21320)
2.	HS and transcription	HSFA2 (AT2G26150, HSFA7a (AT3G51910), HSFA7b (AT3G63350), HSFB1 (AT4G36990),HSFB2a (AT5G62020), HSFB2b (AT4G11660)
3.	Transcription	AZF3 (AT5G43170), DRE2B (AT2G38340), DREB2A (AT5G05410), DREB2B (AT3G11020),HAP5B (AT1G56170), MYB7 (AT2G16720), RHL41 (AT5G59820), SDG37 (AT2G17900)
4.	Apoptosis	BI1 (AT5G47120), BAG6 (AT2G46240)
5.	Cell division	CDC48 (AT5G03340)
6.	Protein degradation	FTSH4 (AT2G26140), FTSH6 (AT5G15250), PEPTIDASE M48 FAMILY PROTEIN (AT5G51740)
7.	Protein synthesis	eEF1Ba1 (AT5G12110)
8.	Metabolism	APX2 (AT3G09640), GOLS1 (AT2G47180), GOLS2 (AT1G56600), MBF1C (AT3G24500), MIPS2 (AT2G22240)
9.	Transport	TIL1 (AT5G58070)

Representative genes of different functions are shown by Thimm et al.[28]

1.4 IMPACT OF HIGH NIGHT-TIME TEMPERATURE ON PLANTS

Soybean (*Glycine max*) has been extensively studied to determine the effect of HNT on plants. A study revealed that HNT causes the reduction of biomass due to a decrease in main stem growth and increased leaf and branch growth than other organs. Also, the leaf area and leaf weight were worked out to determine its relationship with HNT, showing inhibition of leaf growth.[29] In other studies, a decrease in seed size and a rise in the flowers and pods number were observed as there was an increase in night temperature, thus the seed weight per plant was unaffected by night temperature.[30]

The respiration rate in response to global warming increases as such process consumes a larger portion of total photosynthates.[31] Growth, maintenance, and ion uptake are the major functions attributed to the respiration process that can also be analyzed to see its variation concerning environmental conditions.[32,33] Maintenance respiration involves the turnover of proteins and lipids and maintenance of ion concentration gradients across membranes[34] and is the most susceptible to environmental changes.[35] As the crop yield is affected, HNTs are considered to be adverse.[30] As reported by researchers, a high respiration rate triggers the production of reactive oxygen species (ROS) in many plants.[36] To cope with the hazardous effect of these reactive species the enzymatic and nonenzymatic antioxidants are produced under normal physiological conditions. However, stress conditions can shoot up the antioxidant levels, leading to cell damage,[37] thus decreasing membrane thermal stability (MTS), and ultimately disrupting water, ion, and organic-solute movement across plant membranes. This, in turn, affects the process of carbon production, consumption, transport, and accumulation.[38] Studies to measure electrolyte leakage from tissues, such as leaves, subjected to stresses such as drought,[39] heat,[40] and freezing[41] to evaluate the damage to membranes is by examining MTS. The integrity and functions of biomembrane are susceptible to high temperature, as it leads to changes in the membrane proteins' tertiary and quaternary structures through the enhanced permeability of membranes as noted by the electrolytes losses. Membranes are moving montage of proteins and lipids. Lipids stumble between monolayers diffuse within the plane of a monolayer and rotate about their pivots, with their acyl chains also revolving around C–C bonds. Although, much of the heat sensing occurred through protein unfolding.[42] Yield performance in wheat (*Triticum esculentum* L.) and MTS were directly related, under heat-stressed conditions.[43] In the case of rice which is the staple food for the world's population, a decrease of 10% in rice grain yield has been observed due to a 1°C rise in night-time temperature.[3,44] A study on *Arabidopsis* showed, for the first time, the relationship between the major regulators of HSR and plant growth and development. HSFA1s is not only involved in the regulation of the size of seeds but also in the morphology of plants.[45]

1.5 CROP IMPROVEMENT TO COMBAT HEAT STRESS

To decipher the mechanisms and signaling pathways of how the transcriptional, posttranscriptional, translational, and posttranslational processes regulate the plant response(s) to abiotic stress including fluctuating temperatures, the field of "omics" has played an important role by providing many opportunities and strategies.[46,47] DNA is the root of all molecular mechanisms related to HS tolerance in plants and restrains several HS-responsive genes in their genome. A huge number of potential genes with pivotal roles in HS responses have been recognized using whole genome sequence data and genome-wide expression studies. Transcription products (mRNAs) in the genome, make their transcriptome (transcriptomics) and later on translate into the functional proteins (causes for stress tolerance) called proteome (proteomics). In response to developmental and environmental stimuli, plants undergo posttranscriptional regulation by nonprotein-coding small RNAs or microRNAs (miRNAs).[48] Over-expression of mi-RNA-resistant target genes will help to overcome posttranscriptional gene silencing, and thus may lead to a better expression of the induced trait in novel transgenic plants. Understanding the functions of small RNAs in transcriptome homeostasis, cellular tolerance, phenological and developmental flexibility, and recovery under HS will help in genetic engineering of stress tolerance. Also, microarray technology may used for the systematic expression analysis of (transcriptome) large numbers of genes that are induced or repressed by heat shock.[45,49] Certain microarray studies in *Arabidopsis* lacking the *APX* gene, however, noted that certain HSPs are expressed under other stresses also. However, the HSPs are expressed under HS.[50]

1.6 CROP BIOTECHNOLOGY IN GLOBAL CLIMATE CHANGE

There have been extensive studies on advances in the understanding of crop–environment crosstalk at the physiological, morphological, molecular, and biochemical levels, with their significance to biotechnological crop improvement. In the field of biotechnology, scientists are implementing multi-dimension research tools and technologies to figure out, enforce, and thereby influence the genetic makeup of organisms for use in diverse disciplines of agriculture. Transgenic crops can grow faster and development time being reduced from years to months as compared to conventional breeding.[51] Other benefits include an increase in crop yields and an improvement in the control of insects, pests, disease, and climate resistance. However, these crops face criticism and are also supported by fanatics and thus, sparking many debates globally. Therefore, new climate-resilient crops characterized by drought and HS tolerance are required for sustainable food production. However, unpredictable CC poses a challenge to crop scientists, to select the most appropriate traits for crop improvement under limited time and resources.

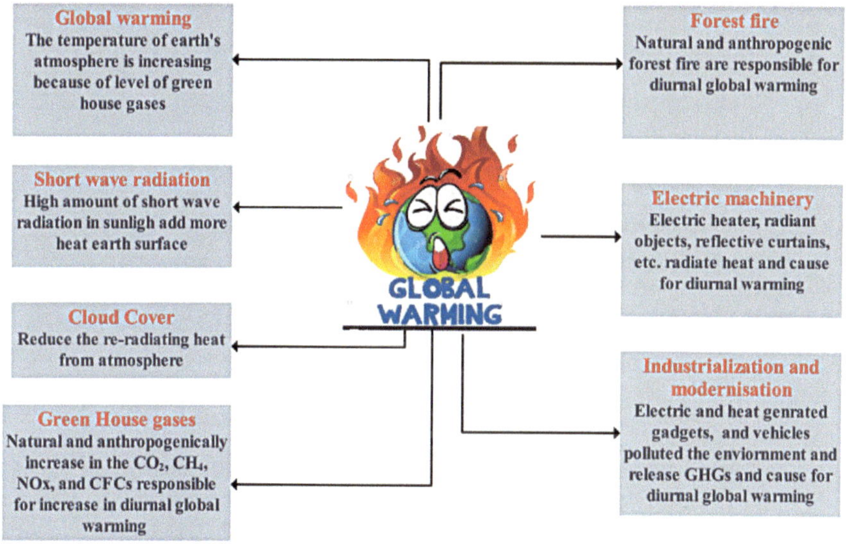

FIGURE 1.1 Natural and anthropogenic factors responsible for diurnal global warming.

1.7 CONCLUSIONS

Elevating global temperatures and unpredictably changing climatic conditions have led to increased night-time temperatures worldwide, thus affecting plants in several ways. The physiological, biochemical, and anatomical changes and changes at the proteomic level are taking place in plants in response to such HS conditions. Ultimately the yield of the plants matters for fulfilling human needs, which is badly affected by a range of abiotic and biotic stresses. We need to understand the existing mechanism in response to those changes, track them, and design such plants, to have the ability to tolerate and flourish even under stressed conditions without getting affected at physiological, biochemical, and anatomical aspects; hence no effect on yield should persist. With the aid of genomic resource databases and "omics" technology, it has now become possible to use a wide range of tools available to manipulate plant genomes for incorporating the desired trait in the plant of interest without altering their good qualities. Hence, crop biotechnology is improving day by day toward sustainable plant adaptations to these CCs for ensuring global food security.

KEYWORDS

- **dehydration**
- **global temperature**
- **humidity**
- **leaf area index**
- **precipitation**

REFERENCES

1. Asseng, S.; Ewert, F.; Martre, P.; Rötter, R. P.; Lobell, D. B.; Cammarano, D.; Kimball, B. A.; Ottman, M. J.; Wall, G. W.; White, J. W.; Reynolds, M. P. Rising Temperatures Reduce Global Wheat Production. *Nat. Clim. Change* **2015,** *5*, 143–147.
2. IPCC. Intergovernmental Panel on Climate Change Fourth Assessment Report: Climate Change. *World Meteorolog.Org.*, 2014.
3. Peng, S.; Huang, J.; Sheehy, J. E.; Laza, R. C.; Visperas, R. M.; Zhong, X.; Centeno, G. S.; Khush, G. S.; Cassman, K. G. Rice Yields Decline with Higher Night Temperature from Global Warming. *Proc. Natl. Acad. Sci.* **2004,** *101* (27), 9971–9975.NOAA. Annual Global Climate Report, 2018. https://www.ncdc.noaa.gov/sotc/global/201813
4. Ming, T.; Liu, W.; Caillol, S. Fighting Global Warming by Climate Engineering: Is the Earth Radiation Management and the Solar Radiation Management Any Option for Fighting Climate Change? *Renewab. Sustain. Energy Rev.* **2014,** *31*, 792–834.

5. Zandalinas, S. I.; Fritschi, F. B.; Mittler, R. Global Warming, Climate Change, and Environmental Pollution: Recipe for a Multifactorial Stress Combination Disaster. *Trends Plant Sci.* **2021,** *26* (6), 588–599.
6. Dessens, J.; Bucher A. Changes in Maximum and Minimum Temperatures at the Pic du Midi in Relation with Humidity and Cloudiness. *Atmosp. Res.* **1995,** *37* (1–3), 147–162.
7. Boucher, O.; Randall, D.; Artaxo, P.; Bretherton, C.; Feingold, G.; Forster, P.; Kerminen, V. M.; Kondo, Y.; Liao, H.; Lohmann, U.; Rasch, P. Clouds and Aerosols. In: *Climate Change 2013: The Physical Science Basis. Contribution of Working Group I to the Fifth Assessment Report of the Intergovernmental Panel on Climate Change*; Cambridge University Press, 2013; pp 571–657.
8. Hartmann, D. L.; Ockert-Bell, M. E.; Michelsen, M. L. The Effect of Cloud Type on Earth's Energy Balance: Global Analysis. *J. Clim.* **1992,** *5* (11), 1281–1304.
9. Budyko, M. I. The Effect of Solar Radiation Variations on the Climate of the Earth. *Tellus* **2016,** *21* (5), 611–617.
10. Kukla, G.; Karl, T. R. Night-Time Warming and the Greenhouse Effect. *Environ. Sci. Technol.* **1993,** *27*, 1468–1474.
11. Speights, C.; Wolff. C. L.; Barton, M. E.; Barton, B. T. Why and How to Create Night Time Warming Treatments for Ecological Field Experiments. *Yale J. Biol. Med.* **2018,** *91*, 471–480.
12. Laude, H. H. Diurnal Cycle of Heat Resistance in Plants. *Science* **1939,** *89*, 556–557.
13. Dodd, A. N.; Jakobsen, M. K.; Baker, A. J.; Telzerow, A.; Hou, S. W.; Laplaze, L.; Barrot, L.; Poethig, R. S.; Haseloff, J.; Webb, A. A. R. Time of Day Modulates Low-Temperature Ca Signals in Arabidopsis. *Plant J.* **2006,** *48*, 962–973.
14. Lee, C. M.; Thomashow, M. F. Photoperiodic Regulation of the C-Repeat Binding Factor (CBF) Cold Acclimation Pathway and Freezing Tolerance in *Arabidopsis thaliana. Proc. Natl. Acad. Sci. USA* **2012**, *109*, 15054–15059.
15. Wu, C.; Heat Shock Transcription Factors: Structure and Regulation. *Ann. Rev. Cell Develop. Biol.* **1995,** *11*, 441–469.
16. Nover, L.; Bharti K.; Doring P.; Mishra S. K.; Ganguli A.; Scharf K. D.; Arabidopsis and the Heat Stress Transcription Factor World: How Many Heat Stress Transcription Factors Do We Need? *Cell Stress Chaperones* **2001,** *6*, 177–189.
17. Shiu, S. H.; Shih, M. C.; Li, W. H. Transcription Factor Families Have Much Higher Expansion Rates in Plants Than in Animals. *Plant Physiol.* **2005,** *139*, 18–26.
18. Wang, C.; Zhang, Q.; Shou, H. X. Identification and Expression Analysis of OsHsfs in Rice. *J. Zhejiang Univ. Sci. B* **2009,** *10*, 291–300.
19. Åkerfelt, M.; Morimoto R. I.; Sistonen L. Heat Shock Factors: Integrators of Cell Stress, Development and Lifespan. *Nat. Rev. Mol. Cell Biol.* **2010,** *11*, 545–555.
20. Hayhoe, K.; Cayan, D.; Field, C. B.; Frumhoff, P. C.; Maurer, E. P.; Miller, N. L.; Moser, S. C.; Schneider, S. H.; Cahill, K. N.; Cleland, E. E.; Dale, L. Emissions Pathways, Climate Change, and Impacts on California. *Proc. Natl. Acad. Sci.* **2004,** *101*,12422–12427.
21. Neill, S. J.; Desikan, R.; Clarke, A.; Hurst R. D.; Hancock J. T. Hydrogen Peroxide and Nitric Oxide as Signalling Molecules in Plants. *J. Exp. Bot.* **2002,** *53*, 1237–1247.

22. Gong, M.; van der Luit, A. H.; Knight, M. R.; Trewavas, A. J. Heat-Shock-Induced Changes in Intracellular Ca^{2+} Level in Tobacco Seedlings in Relation to Thermotolerance. *Plant Physiol.* **1998,** *116*, 429–437.
23. Saidi, Y.; Finka, A.; Muriset, M.; Bromberg, Z.; Weiss, Y. G.; Maathuis, F. J. M.; Goloubinoff, P. The Heat Shock Response in Moss Plants Is Regulated by Specific Calcium-Permeable Channels in the Plasma Membrane. *Plant Cell* **2009,** *21*, 2829–2843.
24. Liu, H. T.; Gao, F.; Li, G. L.; Han, J. L.; Liu, D. L.; Sun, D. Y.; Zhou, R. G. The Calmodulin-Binding Protein Kinase 3 Is Part of Heat Shock Signal Transduction in *Arabidopsis thaliana. Plant J.* **2008,** *55*, 760–773.
25. Schramm, F.; Larkindale, J.; Kiehlmann, E.; Ganguli, A.; Englich, G.; Vierling, E.; von Koskull-Doring, P. A Cascade of Transcription Factor DREB2A and Heat Stress Transcription Factor HsfA3 Regulates the Heat Stress Response of *Arabidopsis. Plant J.* **2008,** *53*, 264–274.
26. Yoshida, T.; Sakuma, Y.; Todaka, D.; Maruyama, K.; Qin, F.; Mizoi, J.; Kidokoro, S.; Fujita, Y.; Shinozaki, K.; Yamaguchi-Shinozaki, K. Functional Analysis of an Arabidopsis Heat-Shock Transcription Factor HsfA3 in the Transcriptional Cascade ownstream of the DREB2A Stress-Regulatory System. *Biochem. Biophy. Res. Commu.* **2008,** *368*, 515–521.
27. Thimm, O.; Blasing, O.; Gibon, Y.; Nagel, A.; Meyer, S.; Kruger, P.; Selbig, J.; Muller, L. A.; Rhee, S. Y.; Stitt, M. MAPMAN: A User-Driven Tool to Display Genomics Data Sets onto Diagrams of Metabolic Pathways and Other Biological Processes. *Plant J.* **2004,** *37*, 914–939.
28. Lin, T.; Okamoto, Y.; Nagasaki, Y.; Shiraiwa, T. The Influence of High Night Temperature on Yield and Physiological Attributes of Soybean cv. Fukuyutaka. *Plant Prod. Sci.* **2021,** *24* (3), 267–278.
29. Zheng, S. H.; Nakamoto, H.; Yoshikawa, K.; Furuya, T.; Fukuyama, M. Influence of High Night Temperature on Flowering and Pod Setting in Soybean. *Plant Prod. Sci.* **2002,** *5*, 215–218.
30. Paembonan, S. A.; Hagihara, A.; Hozumi, K. Long-Term Respiration In Relation to Growth and Maintenance Processes of the Above Ground Parts of a Hinoki Forest Tress. *Tree Physiol.* **1992,** *10*, 101–110.
31. Lambers, H. Respiration in Tack Plants and Tissues: Its Regulation and Dependence on Environmental Factors, Metabolism and Invaded Organism. In: *Higher Plant Cell Respiration: Encyclopedia of Plant Physiology*; Douce, R., Day, D. A., Eds.; New Ser., Vol. 18; Springer: Berlin, 1985.
32. Amthor, J. S. Evolution and Applicability of a Whole Plant Respiration Model. *J. Theor. Biol.* **1986,** *122*, 473–490.
33. Penning de Vries, F. W. T.; The Cost of Maintenance Process in the Cell. *Ann. Bot.* **1975,** *39*, 77–92.
34. Ryan, M. G. Effects of Climate Change on Plant Respiration. *Ecol. App.* **1991,** *1* (2), 157–167.
35. McDonald, A. E.; Vanlerberghe, G. C. Alternative Oxidase and Plastoquinol Terminal Oxidase in Marine Prokaryotes of the Sargasso Sea. *Gene* **2005,** *349*, 15–24.

36. Kreiner, M.; Harvey, L. M.; McNeil, B. Oxidative Stress Response of a Recombinant *Aspergillusniger* to Exogenous Menadione and H_2O_2 Addition. *Enzyme Microb. Technol.* **2002,** *30,* 346–353.
37. Christiansen, M. N. The Physiology of Plant Tolerance to Temperature Extremes. In: *Crop Tolerance to Suboptimal Land Conditions*; Jung, G. A., Ed.; ASA: Madison, 1978; pp 173–191.Blum, A.; Ebercon, A. Cell Membrane Stability as a Measure of Drought and Heat Tolerance in Wheat. *Crop Sci.* **1981,** *21,* 43–47.
38. Sullivan, C. Y. Mechanisms of Heat and Drought Resistance in Grain Sorghum and Methods of Measurement. In: *Sorghum in the Seventies*; Rao, N. G. P., House, L. R., Eds.; Oxford and IPH: New Delhi, 1972; pp 247–264. .
39. Dexter, S. T.; Evaluation of Crop Plants for Winter Hardiness. *Adv. Agron.* **1956,** *8,* 203–209.
40. Hemantaranjan, A.; Bhanu, A. N.; Singh, M. N.; Yadav, D. K.; Patel, P. K.; Singh, R.; Katiyar, D. Heat Stress Responses and Thermotolerance. *Adv. Plants Agric. Res.* **2014,** *1* (3), 62–70.
41. Reynolds, M. P.; Balota, M.; Delgado, M. I. B.; Amani, I.; Fisher, R. A. Physiological and Morphological Traits Associated with Spring Wheat Yield Under Hot Irrigated Conditions. *Aust. J. Plant Physiol.* **1994,** *21,* 717–730.
42. Mohammed, A. R.; Tarpley, L. Impact of High Night-time Temperature on Respiration, Membrane Stability, Antioxidant Capacity, and Yield of Rice Plants. *Crop Sci.* **2009,** *49,* 313–322.
43. Liu, H. C.; Liao, H. T.; Charng, Y. Y. The role of class A1 heat shock factors (HSFA1s) in Response to Heat and Other Stresses in Arabidopsis. *Plant Cell Environ.* **2011,** *34,* 738–751.
44. Aprile, A.; Mastrangelo, A. M.; De Leonardis, A. M.; Galiba, G.; Roncaglia, E.; Ferrari, F.; De Bellis, L.; Turchi, L.; Giuliano, G.; Cattivelli, L. Transcriptional Profiling in Response to Terminal Drought Stress Reveals Differential Responses Along the Wheat Genome. *BMC Genom.* **2009,** *10* (1), 1–18.
45. Hasanuzzaman, M.; Nahar, K.; Fujita, M.; Ahmad, P.; Chandna, R.; Prasad, M. N. V.; Ozturk, M. Enhancing Plant Productivity Under Salt Stress: Relevance of Poly-Omics. *Salt Stress in Plants***2013,** 113–156.
46. Chinnusamy, V.; Zhu, J.; Zhou, T.; Zhu, J. K. Small RNAs: Big Role in Abiotic Stress Tolerance of Plants. In *Advances in Molecular Breeding Toward Drought and Salt Tolerant Crops*; Springer: Dordrecht, 2007; pp 223–260.
47. Yeh, C. H.; Kaplinsky, N. J.; Hu, C.; Charng, Y. Y. Some Like It Hot, Some Like It Warm: Phenotyping to Explore Thermotolerance Diversity. *Plant Sci.* **2012,** *195,* 10–23.
48. Pnueli, L.; Liang, H.; Rozenberg, M.; Mittler, R. Growth Suppression, Altered Stomatal Responses and Augmented Induction of Heat Shock Proteins in Cytosolic Ascorbate Peroxidase (Apx1)-Deficient Arabidopsis Plants. *Plant J.* **2003,** *34* (2), 187–203.
49. Stephenson, L. A. Germany's an of Monsanto's Genetically Modified Maize (MON810), a Violation of International Law. *Trade L. Dev.* **2010,** *2* (2), 292–328.

CHAPTER 2

Elevated CO_2 and Night Warming for Plant Growth and Development

DALPAT LAL[1], NEHA SHARMA[2], and YOGESH KHOKHAR[3]

[1]*Department of Agriculture, Jagan Nath University, Jaipur, India*

[2]*Amity Institute of Biotechnology, Amity University, Noida, Uttar Pradesh, India*

[3]*ICAR-Indian Grassland and forage Research Institute, Jhansi, India*

ABSTRACT

Today the world population is continuously increasing and human comfort is accelerating greenhouse gasses and climate change. The accelerating pace of climate change combined with the global population is a major challenge to global food security. The pattern of climate change leads to the development of new races or resistant types of pests and diseases that affect crop production. The climate is continuously changing due to an increase in greenhouse gases and human activity. Effect of climate on all the abiotic and biotic factors which lead to a change in agriculture production. The evaluated CO_2 has a beneficial effect on plant growth and development by enhancing photosynthetic products such as carbohydrates and biomass. Further, it also affects physiological processes such as transpiration, respiration, plant growth regulators, etc. The increasing temperature harms crop production and productivity. Moreover, the night warming has impacts of increased temperature have a greater effect on grain yield than vegetative growth.

Impact of High Night Temperature on Plant Biology: Toward Sustainable Plant Adaptation to Climate Change. Rajesh Kumar Singhal, Raju Bheemanahalli, Saurabh Pandey, and Pratibha M. D. (Eds.)

2.1 INTRODUCTION

In recent years, it has become apparent that temperatures may increase due to global climate change.[1] The concentration of atmospheric carbon dioxide climbed from 300 to 350 ppm throughout this century, and it may double during the following century.[2] According to global climate models, rising carbon dioxide (CO_2) and other greenhouse gas concentrations could result in higher global temperatures. Recent climate model estimates have also revealed that this doubling of CO_2 concentration may result in a rise in global surface air temperature of 1.4–5.8°C.[3] It is critical to determine how CO_2 and air temperature affect the growth and productivity of essential crop plants since they are two of the environmental factors that regulate phenological and physiological processes in plant crops. It is of particular interest to estimate the interactions of these two climate variables because changes in temperature and carbon dioxide concentration are likely to occur concurrently. The reproductive stages of several plants, including tomato, soybean, *Phaseolus vulgaris* L., *Gossypium barbadense* L., and rice, have been proven to be particularly sensitive to heat, which significantly reduces agricultural yield in various climatic zones.[4,5] Several crucial plant processes such as photosynthesis were crucially affected by the interaction of elevated CO_2 and night warming. Therefore, combined studies of the responses to elements alone and their combinations are required to understand the impacts of various changes on plant physiological systems under a changed climate.[6,7]

2.2 CROP GROWTH UNDER ELEVATED CARBON DIOXIDE

Effects of CO_2 have been observed on several plant crops wherein the response was correlated with additional environmental conditions including light, water, temperature, and salinity,[8] especially phosphate and nitrate.[9] Enhanced growth and photosynthesis are typically seen in C_3 plants in response to increased carbon dioxide. However, a variety of acclimatization processes can reduce the effects of increased carbon dioxide concentration on plants' responses.[8,10,11] Changes in photosynthesis and growth in C_4 plants in response to high carbon dioxide would not significantly alter due to the anatomical and biochemical specialization related to the CO_2 mechanism. However, multiple research studies revealed that increased CO_2 had a positive impact on C_4 species' photosynthesis and plant development. C_4 plants are more adapted to high temperatures than C_3 plants since they have greater temperature optima for photosynthesis and growth.[12] It is unknown if photosynthesis and growth in C_4 plants would still be temperature-dependent at higher CO_2 levels than at present.[13] The ideal temperature for photosynthesis in C_3 plants was greater than at ambient CO_2, and high CO_2 stimulated photosynthesis more effectively at higher temperatures.[14] It may be comparable to how temperature and CO_2 interact to affect the photosynthesis of C_3 plants, but further research is needed to determine this.[15]

The mechanisms by which increased CO_2 can alleviate the negative effects of abiotic stressors include higher photosynthesis, improved water usage efficiency as a result of decreased stomatal conductance, elevated antioxidant levels, and their activities, as well as other factors including decreased photorespiration, that cause the reactivity of oxygen species.

2.3 EFFECT OF NIGHT WARMING ON CROPS

Increased night temperature accelerates various components of plant metabolism.[16] This can have an impact on crops by altering the timing and speed of physiological processes such as senescence, respiration, and organ development. High temperatures can accelerate the growth of plants, causing earlier transitions to the next stage.[17] As photosynthesis only takes place during the day and respiration takes place during both day and night times, day and night warming may change the ratio between respiration and photosynthesis.

According to reports 50 % of net photosynthetic carbon fixation is released via plant respiration during the following night. According to Turnbull et al.,[18] changes in night-time minimum temperatures may have a major impact on net plant carbon intake since they improved photosynthetic capability during the subsequent light period through a respiratory-driven decrease in the leaf carbohydrate content. According to studies, day and night warming have different effects on the interaction between the carbon sink and source, which in turn affects how plants function physiologically. Crop responses to temperature depending on the specific optimal temperature for growth, photosynthesis, and yield.[19] If the temperature is below the level necessary for photosynthesis, a slight temperature rise may accelerate plant growth and development; however, if the temperature is already high, an insignificant rise in temperature may hinder crop development and reduce yield.[20] Similarly, how a crop reacts to temperature warming depends on temperature increase and the stage of the crop's development. High temperatures have a negative impact on crop reproduction, particularly flowering and fertilization, and reduce plant biomass. High temperatures reduce plant yield and biomass by lowering photosynthesis and increasing stomatal conductance and transpiration. Plants tend to produce more immense root systems and smaller leaves under higher temperatures to increase water intake and offset leaf water loss.

Furthermore, plant hormones like ethylene, abscisic acid (ABA), and indole-3-acetic acid (IAA) are involved in the responses of plants to temperature warming by balancing transpiration by influencing stomatal conductance. High temperatures decrease IAA production and increase ABA and ethylene evolution. IAA negatively affects plant growth under higher temperatures.

Young et al.[20] observed that, in canola, high temperature reduced pollen germinability and viability, induced fruit abortion, and affected seed production. In annual crops, such as wheat,[21] and peanut (*Arachis hypogaea* L.),[22] due to pollen sterilization, even a short period of high temperature close to the flowering may significantly reduce production. According to Mitchell et al.,[23] shortened phenological phases and consequently less time for accumulation of resources for seed formation may be the cause of the decreased yield in wheat. The quality of the crop can also be affected by high temperatures. High temperatures have been shown to decrease polypeptide content and change the ratios and types of lipids and proteins in cereal grains.[24]

2.4 INTERACTIONS BETWEEN CROP RESPONSES TO INCREASED CO_2 AND NIGHT WARMING

Plant scientists are studying the combined effects of elevated CO_2 and night warming on plants, especially crops, due to concerns about the concentration of atmospheric CO_2 and its correlation to high temperatures, which may influence crop productivity.[25–27] According to Vu et al.,[28] increasing CO_2 increased the rate of CO_2 uptake by leaves during photosynthetic activity, whereas rice and soybean Rubisco concentrations were decreased by CO_2 and night warming. Due to differences in how these crops react to these environmental conditions, C_3 plants may respond differently to future global climate change than other plant species. Some studies studying the same crop species have reported conflicting results in response to high temperatures and increasing CO_2 levels (Table 2.1). The interactive effect of elevated CO_2 and night warming is highlighted in Figure 2.1. The elevated CO_2 and warming effect combined effect on the N cycle was tested in the field and also studied N cycle under ambient and elevated CO_2 concentration and ambient and elevated temperature in the semiarid grassland field. They reported that N was significantly decreased under elevated CO_2 but the sol inorganic N and plant N pool sizes significantly increased with warming. Furthermore, it was also observed that no CO_2 ×Warming interacting effect on soil inorganic N, N pool size, or NO_3 uptake in plants and microbes.[43] The impact of elevated CO_2 concentration and warming on plant nutrient ratio was studied in open conditions. They have studied the plant C, N, and P ratios in response to elevated (CO_2) combined with canopy warming in China for the rice–winter wheat rotation system. The conditions were ambient conditions, elevated (CO_2) (500 ppm), canopy warming (+2 °C), and the combination of elevated (CO_2) and warming was used to study the responses of plant C, N, and P ratios. The results showed that elevated (CO_2) increased the C:N ratio by 8.4–14.3% for both crops and increased the C:P ratio by 11.3% for rice. Also, the results showed the change in C (increasing), N (decreasing), and P (increasing) in rice.[45] A relative effect of increasing the day and night time temperature on tree carbon has been reported in tress species *Populus deltoids*. They have studied respiration during the night and photosynthetic capacity in the day the time in the leaves. The results suggested that elevated night-time temperature enhances the capacity of photosynthesis during light time through a respiratory-driven reduction in leaf carbohydrate concentration. These results suggested that the predicted increases in night-time minimum temperatures may have a significant influence on net plant carbon uptake.[18]

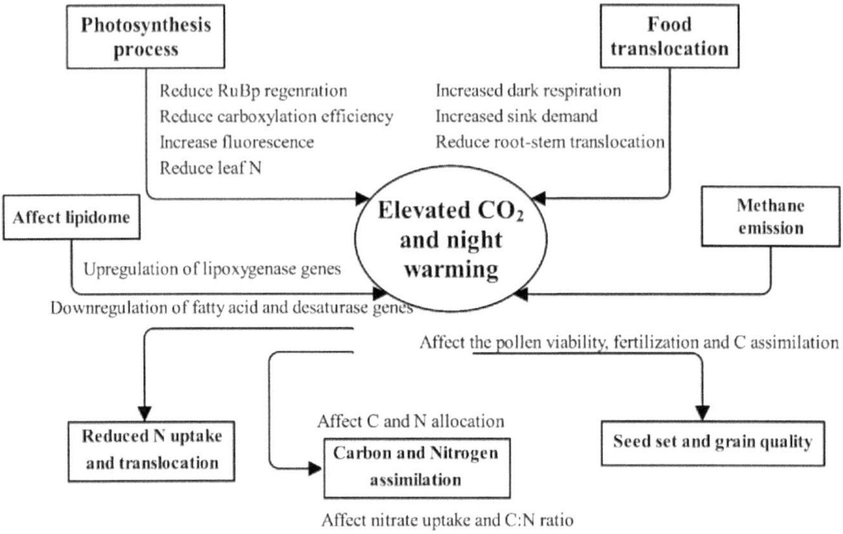

FIGURE 2.1 Highlights the impact of eCO_2 and night warming on crucial plant processes.

TABLE 2.1 Experiment Conducted in Different Crops Using the Night-time Environment.

S. No.	Crops	Environment	References
1	Rice	Free-air CO_2 enrichment and greenhouse	[29, 30]
2	Sweet potato	Greenhouse	[31]
3	Peanut	Greenhouse and soil–plant–atmosphere–research unit	[32, 33]
4	Bean	Controlled-environment chamber and sunlight controlled-environment chamber	[34, 35]
5	Soybean	Growth chamber and sunlight-controlled-environment chamber	[36, 37]
6	Wheat	Open–top chamber and chamber–greenhouse	[38, 39]
7	Cotton	Controlled-environment chamber—soil–plant–atmosphere–research unit	[40]
8	Cowpea	Greenhouse	[41]

These variations could be attributed to various experimental strategies, crop development stages, and other environmental variables that were probably not taken into account throughout the studies. However, some plant processes were significantly influenced by their interactions such as maintaining respiration, photosynthesis, and C and N assimilation highlighted in Table 2.2. The interaction of the elevated carbon dioxide and night warming could be positive, negative, or neutral but they can modify the crucial plant processes significantly in future climate change scenarios.

TABLE 2.2 Effect of Night Warming and Elevated CO_2 on Various Plant Processes.

Interactions	Plants	Plant processes	Mechanism	References
Drought × Night warming × eCO_2	*Calluna vulgaris* and *Deshampsia flexuosa*	Photosystem II performance	Reduce maximal velocity RuBP regeneration, maximal carboxylation velocity of Rubisco and total leaf nitrogen content, and total performance index	[42]
Night warming ×eCO_2	*Populus deltoides*	Photosynthesis	Increased dark respiration, increased nonstructural carbohydrates (increased sink demand), and increased photosynthetic rate	[18]
Drought × Night warming × eCO_2	*C. vulgaris*	N cycle	No significant effect on soil organic N, N pool size and NO_3^- uptake	[43]
Night warming × eCO_2	*Oryza sativa* L.	C assimilation, N absorption, and their allocation	Decreased C concentration in the leaves, and increased N allocation.	[44]
Night warming× eCO_2	*O. sativa* L.	C:N ratio	Night warming offset the effect of eCO_2.	[45]
Warming× eCO_2	*Coffea arabica* L. (cv. Icatu and IPR108) and *C. canephora* cv. Conilon CL153	Lipid profile adjustment	Upregulation of LOX5A and LOX5B lipoxygenase gene, downregulation of fatty acid and desaturase (FAD) gene.	[46]
Warming × eCO_2	*Solanum lycopersicum*	N uptake and assimilatory protein	Reduce N uptake by root, total protein concentration in roots (inhibit N assimilation), and reduce shoot N (reduce translocation)	[47]
Warming × eCO_2	Rice paddy soil	Methane emission	Reduce the stimulatory effect of CO_2 on methane emission	[48]
Warming× eCO_2	*O. sativa* L.	Seed set and grain quality	High yield due to high spikelet number, and increased broken grains	[49]

2.5 CONCLUSION

Climate has a significant impact on the world's food production and is the primary determinant of agricultural productivity. Agriculture industries are particularly vulnerable to climate change because of factors such as rising climate factors such as CO_2 concentrations and night warming can limit productivity and endanger the safety of the world's food supply by causing heat or cold stress. Because of the night warming, the impacts of increased temperature have a greater effect on grain yield than on vegetative development. However, reports suggested that night warming and elevated CO_2 act as synergistic and antagonistic during the different plant processes and affect plant growth development. Therefore, it is crucial to carry on more studies on the crosstalk of night warming and elevated carbon dioxide for a better understanding of their role in plant growth and regulation in future climate change scenarios.

KEYWORDS

- **greenhouse gasses**
- **climate change**
- **food security**
- **elevated CO2**
- **night warming**

REFERENCE

1. IPCC. Summary for Policymakers. In *Climate Change: The Physical Science Basis. Contribution of Working Group I to the Sixth Assessment Report of the Intergovernmental Panel on Climate Change*; Masson-Delmotte,V., Zhai, P., Pirani, A., Connors, S. L., Pe´an, C., Berger, S., Caud, N., Chen, Y., Goldfarb, L., Gomis, M. I., Huang, M., Leitzell, K., Lonnoy, E., Matthews, J. B. R., Maycock, T. K., Waterfield, T., Yelekci, O., Yu, R., Zhou, B., Eds.; Cambridge University Press: Cambridge, 2021; pp 17–18.
2. Prentice, I. C.; Farquhar, G. D.; Fasham, M. J. R.; Goulden, M. L.; Heimann, M.; Jaramillo, V. J.; Kheshgi, H. S.; Qu´er´e, C. L.; Scholes, R. J.; Wallace, D. W. R.; The Carbon Cycle and Atmospheric Carbon Dioxide. In: *Climate Change: The*

Scientific Basis; Johnson, C. A., Ed.; Cambridge University Press: Cambridge, 2001; pp 183–237.
3. Cubasch, U.; Meehl, G. A.; Boer, G. J.; Stouffer, R. J.; Dix, M.; Noda, A.; Senior, C. A.; Raper, S.; Yap, K. S. Projections of Future Climate Change. In: *Climate Change: The Scientific Basis*; Johnson, C. A., Ed.; Cambridge University Press: Cambridge, 2001; pp 525–582 (Chapter 9).
4. Kimball, B. A. Carbon Dioxide and Agricultural Yield: An Assemblage and Analysis of 430 Prior Observations. *Agron. J.* **1983,** *75*, 779–788.
5. Poorter, H.; Van Berkel, Y.; Baxter, R.; Den Hertog, J.; Dijkstra, P.; Gifford, R. M.; Griffin, K. L.; Roumet, C.; Roy, J.; Wong, S. C. The Effects of Elevated CO_2 on the Chemical Composition and Construction Costs of Leaves of 27 C_3 Species. *Plant Cell Environ.* **1997,** *20*, 472–482.
6. Williams, M.; Shewry, P. R.; Harwood, J. L. The Influence of the 'Greenhouse Effect' on Wheat (*Triticum aestivum* L.) Grain lipids. *J. Exp. Bot.* **1994,** *45*, 1379–1385.
7. Sabagh, A. E.; Hossain, A.; Islam, M. S.; Iqbal, M. A.; Raza, A.; Karademir, Ç.; Karademir, E.; Rehman, A.; Rahman, M. A.; Singhal, R. K.; Llanes, A. Elevated CO_2 Concentration Improves Heat-Tolerant Ability in Crops. In: *Abiotic Stress in Plants*; IntechOpen 2020.
8. Stitt, M. Rising CO_2 Levels and Their Potential Significance for Carbon Flow in Photosynthetic Cells. *Plant Cell Environ.* **1991,** *14*, 741–762.
9. Delgado, E.; Mitchell, R. A. C.; Parry, M. A. J.; Driscoll, S. P.; Mitchell, V. J.; Lawlor, D. W. Interacting Effects of CO_2 Concentration, Temperature and Nitrogen Supply on the Photosynthesis and Composition of Winter Wheat Leaves. *Plant, Cell Environ.* **1994,** *17*, 1205–1213.
10. Long, S. P. Modification of the Response of Phtosynthetic Productivity to Rising Temperature by Atmospheric Concentrations: Has Its Importance Been Underestimated? *Plant, Cell Environ.* **1991,** *14*, 729–739.
11. McKee, I. F.; Woodward, F. I.; CO_2 Enrichment Responses of Wheat: Interactions with Temperature, Nitrate and Phosphate. *New Phytol.* **1994,** *127*, 447– 453.
12. Moore, B. D.; Cheng, S. H.; Sims, D.; Seemann, J. R. The Biochemical and Molecular Basis for Photosynthetic Acclimation to Elevated Atmospheric CO_2. *Plant Cell Environ.* **1999,** *22*, 567–582.
13. Ghannoum, O.; von Caemmerer, S.; Ziska, L. H.; Conroy, J. P. The Growth Response of C_4 Plants to Rising Atmospheric CO_2 Partial Pressure: A Reassessment. *Plant, Cell Environ.* **2000,** *23*, 931–942.
14. Kirschbaum, M. The Sensitivity of C_3 Photosynthesis to Increasing CO_2 Concentration: A Theoretical Analysis of Its Dependence on Temperature and Background CO_2 Concentration. *Plant, Cell Environ.* **1994,** *17*, 747–754.
15. Morison, J. I. L.; Lawlor, D. W. Interactions Between Increasing CO_2 Concentration and Temperature on Plant Growth. *Plant, Cell Environ.* **1999,** *22*, 659–682.
16. Larcher, W. *Physiological Plant Ecology*, 4th edn.; Springer-Verlag: Berlin Heidelberg, 2003.
17. Badeck, F. W.; Bondeau, A.; B¨ottcher, K.; Doktor, D.; Lucht, W.; Schaber, J.; Sitch, S. Responses of Spring Phenology to Climate Change. *New Phytol.* **2004,** *162*, 295–309.

18. Turnbull, M.; Murthy, R.; Griffin, K. The Relative Impacts of Daytime and Night-Time Warming on Photosynthetic Capacity in *Populus deltoides*. *Plant Cell Environ* **2002,** *25*, 1729–1737.
19. Polowick, P. L.; Sawhney, V. K. High Temperature Induced Male and Female Sterility in Canola (*Brassica napus* L.). *Ann. Bot.* **1988,** *62*, 83–86.
20. Young, L. W.; Wilen, R. W.; Bonham-Smith, P. C. High Temperature Stress of *Brassica napus* During Flowering Reduces Micro- and Megagametophyte Fertility, Induces Fruit Abortion, and Disrupts Seed Production. *J. Exp. Bot.* **2004,** *55*, 485–495.
21. Wheeler, T. R.; Craufurd, P. Q.; Ellis, R. H.; Porter, J. R.; Prasad, P. V. V. Temperature Variability and the Yield of Annual Crops. *Agric. Ecosyst. Environ.* **2000,** *82*, 159–167.
22. Challinor, A. J.; Wheeler, T. R.; Craufurd, P. Q.; Slingo, J. M. Simulation of the Impact of High Temperature Stress on Annual Crop Yields. *Agric. Forest Met.* **2005,** *135*, 180–189.
23. Mitchell, R. A. C.; Mitchell, V. J.; Driscoll, S. P.; Franklin, J.; Lawlor, D. W. Effects of Increased CO_2 Concentration and Temperature on Growth and Yield of Winter Wheat at Two Levels of Nitrogen Application. *Plant Cell Environ.* **1993,** *16*, 521–529.
24. Lawlor, D. W. Plant Responses to Climate Change: Impacts and Adaptation. In: *Plant Responses to Air Pollution and Global Change*; Omasa, K., Nouchi, I., De Kok, L. J., Eds.; Springer-Verlag: Tokyo, 2005; pp 81–88.
25. Idso, S. B.; Kimball, B. A.; Anderson, M. G.; Mauney, J. R. Effect of Atmospheric CO_2 Enrichment on Plant Growth: The Interactive Role of Air Temperature. *Agric. Ecosyst. Environ.* **1987,** *20*, 1–10.
26. Rawson, H. M. Plant Responses to Temperature Under Conditions of Elevated CO_2. *Aust. J. Bot.* **1992,** *40*, 473–490.
27. Morison, J. I. L.; Lawlor, D. W. Interactions Between Increasing CO_2 Concentration and Temperature on Plant Growth. *Plant Cell Environ.* **1999,** *22*, 659–682.
28. Vu, J. C. V.; Allen, L. H.; Jr, Boote, K. J.; Bowes, G.; Effects of Elevated CO_2 and Temperature on Photosynthesis and Rubisco in Rice and Soybean. *Plant Cell Environ.* **1997,** *20*. 68–76.
29. Borjigidai, A.; Hikosaka, K.; Hirose, T.; Hasegawa, T.; Okada, M.; Kobayashi, K.; Seasonal Changes in Temperature Dependence of Photosynthetic Rate in Rice Under a Free-Air CO_2 Enrichment. *Ann Bot.* **2006,** *97*, 549–557.
30. Ziska, L. H.; Manalo, P. A. Ordonez RA Intraspecific Variation in the Response of Rice (*Oryza sativa* L.) to Increased CO_2 and Temperature: Growth and Yield Response of 17 Cultivars. *J. Exp. Bot.* **1996,** *47*, 1353–1359.
31. Cen, Y. P.; Sage, R. E. The Regulation of Rubisco Activity in Response to Variation in Temperature and Atmospheric CO_2 Partial Pressure in Sweet Potato. *Plant Physiol.* **2005,** *139*, 979–990.
32. Vu, J. C. V. Acclimation of Peanut (*Arachis hypogaea* L.) Leaf Photosynthesis to Elevated Growth CO_2 and Temperature. *Environ. Exp. Bot.* **2005,** *53*, 85–95.
33. Prasad, P. V. V.; Allen, L. H.; Jr, Boote, K. J. Crop Responses to Elevated Carbon Dioxide and Interaction with Temperature: Grain Legumes. *J. Crop Improv.* **2005,** *13*, 113–155.

34. Cowling, S. A.; Sage, R. F. Interactive Effects of Low Atmospheric CO_2 and Elevated Temperature on Growth, Photosynthesis and Respiration in *Phaseolus vulgaris*. *Plant Cell Environ.* **1998,** *21*, 427–435.
35. Prasad, P. V. V.; Boote, K. J.; Allen, L. H.; Jr, Thomas, J. M. G. Effects of Elevated Temperature and Carbon Dioxide on Seed-Set and Yield of Kidney Bean (*Phaseolus vulgaris* L.). *Glob. Change Biol.* **2002,** *8*, 710–721.
36. Sionit, N.; Strain, B. R.; Flint, E. P. Interaction of Temperature and CO_2 Enrichment on Soybean: Photosynthesis and Seed Yield. *Can. J. Plant Sci.* **1987,** *67*, 629–636.
37. Thomas, J. M. G.; Boote, K. J.; Allen, L. H.; Jr, Gallo-Meagher, M.; Davis, J. M. Elevated Temperature and Carbon Dioxide Effects on Soybean Seed Composition and Transcript Abundance. *Crop Sci.* **2003,** *43*, 1548–1557.
38. Hakala, K. Growth and Yield Potential of Spring Wheat in a Simulated Changed Climate with Increased CO_2 and Higher Temperature. *Eur. J. Agron.* **1998,** *9*, 41–52.
39. Mitchell, R. A. C.; Mitchell, V. J.; Driscoll, S. P.; Franklin, J.; Lawlor, D. W. Effects of increased CO_2 Concentration and Temperature on Growth and Yield of Winter Wheat at Two Levels of Nitrogen Application. *Plant Cell Environ.* **1993,** *16*, 521–529.
40. Reddy, K. R.; Kakani, V. G.; Prasad, P. V. V. Crop responses to Elevated Carbon Dioxide and Interactions with Temperature: Cotton. *J. Crop Improv.* **2005,** *13*, 157–191.
41. Ahmed, F. E.; Hall, A. E.; Madore, M. A. Interactive Effects of High Temperature and Elevated Carbon Dioxide Concentration on Cowpea [*Vigna unguiculata* (L.) Walp.]. *Plant Cell Environ.* **1993,** *16*, 835–842.
42. Albert, K. R.; Mikkelsen, T. N.; Michelsen, A.; Ro-Poulsen, H.; van der Linden, L. Interactive Effects of Drought, Elevated CO_2 and Warming on Photosynthetic Capacity and Photosystem Performance in Temperate Heath Plants. *J. Plant Physiol.* **2011,** *168* (13), 1550–1561.
43. Dijkstra, F. A.; Blumenthal, D.; Morgan, J. A.; Pendall, E.; Carrillo, Y.; Follett, R. F. Contrasting Effects of Elevated CO_2 and Warming on Nitrogen Cycling in a Semiarid Grassland. *New Phytol.* **2010,** *187* (2), 426–437.
44. Cheng, W.; Sakai, H.; Yagi, K.; Hasegawa, T. Combined Effects of Elevated [CO_2] and High Night Temperature on Carbon Assimilation, Nitrogen Absorption, and the Allocations of C and N by Rice (*Oryza sativa* L.). *Agric. Forest Meteorol.* **2010,** *150* (9), 1174–1181.
45. Wang, J.; Liu, X.; Zhang, X.; Li, L.; Lam, S. K.; Pan, G. Changes in Plant C, N and P Ratios Under Elevated [CO_2] and Canopy Warming in a Rice-Winter Wheat Rotation System. *Sci. Rep.* **2019,** *9* (1), 1–9.
46. Scotti-Campos, P.; Pais, I. P.; Ribeiro-Barros, A. I.; Martins, L. D.; Tomaz, M. A.; Rodrigues, W. P.; Campostrini, E.; Semedo, J. N.; Fortunato, A. S.; Martins, M. Q.; Partelli, F. L. Lipid Profile Adjustments May Contribute to Warming Acclimation and to Heat Impact Mitigation by Elevated [CO_2] in Coffea spp. *Environ. Exp. Bot.* **2019,** *167*, p.103856.
47. Jayawardena, D. M.; Heckathorn, S. A.; Bista, D. R.; Mishra, S.; Boldt, J. K.; Krause, C. R. Elevated CO_2 Plus Chronic Warming Reduce Nitrogen Uptake and Levels or Activities of Nitrogen-Uptake and-Assimilatory Proteins in Tomato Roots. *Physiol. Planta.* **2017,** *159* (3), 354–365.

48. Cheng, W.; Sakai, H.; Hartley, A.; Yagi, K.; Hasegawa, T. Increased Night Temperature Reduces the stimulatory Effect of Elevated Carbon Dioxide Concentration on Methane Emission from Rice Paddy Soil. *Glob. Chang. Biol.* **2008,** *14* (3), 644–656.
49. Madan, P.; Jagadish, S. V. K.; Craufurd, P. Q.; Fitzgerald, M.; Lafarge, T. Wheeler, T. R. Effect of Elevated CO_2 and High Temperature on Seed-Set and Grain Quality of Rice. *J. Exp. Bot.* **2012,** *63* (10), 3843–3852.

CHAPTER 3

Plant Growth and Development Under High Night Temperatures

BASANT KUMAR DADARWAL[1], NITIN YADAV[1], JYOTI CHAUHAN[2], and RAJESH KUMAR SINGHAL[3]

[1]*Institute of Agricultural Sciences, Banaras Hindu University, Varanasi, Uttar Pradesh, India*

[2]*Noida International University, Noida, Uttar Pradesh, India*

[3]*ICAR- Indian Grassland and Forage Research Institute Jhansi, U.P., Uttar Pradesh, India*

ABSTRACT

Climate change is expected to increase global temperatures, including night temperatures, which could have a significant impact on plant biology. High night temperatures (HNTs) have become one of the critical factors of climate change, which affects plant growth and development. Earlier studies suggest that HNT can disrupt the balance between photosynthesis and respiration, reduce stomatal conductance (g_s), alter hormone balance, and induce oxidative stress in plants. These effects may lead to reduced plant growth, yield, and quality. However, some plant species have been shown to have the ability to adapt to HNTs through changes in gene expression, metabolic pathways, and structural adaptations. This review highlights the current understanding of the impact of HNTs on plant biology and identifies strategies for sustainable plant adaptation to climate change. Understanding the response of plant HNTs is critical for developing effective strategies to mitigate the effects of climate change

Impact of High Night Temperature on Plant Biology: Toward Sustainable Plant Adaptation to Climate Change. Rajesh Kumar Singhal, Raju Bheemanahalli, Saurabh Pandey, and Pratibha M. D. (Eds.)

on agricultural productivity and food security. Furthermore, this chapter highlights the strategies used by plants to adapt to HNTs and discusses the challenges in sustaining plant adaptation to climate change. In conclusion, this chapter emphasizes the importance of understanding the effects of HNTs on plant biology and suggests sustainable solutions to mitigate the negative impact of climate change on plant growth and development.

3.1 INTRODUCTION

Climate change is one of the major challenges faced by the world today, and its impact on plant biology is a matter of great concern. As temperatures rise, plants experience various physiological and biochemical changes that can have adverse effects on their growth and development.[1] One of the critical impacts of climate change on plant biology is the rise in high night temperatures (HNTs). This has been linked to reduced photosynthesis, increased respiration rates, and reduced biomass accumulation.[2] However, plants have evolved several mechanisms to cope with HNTs and ensure their survival. This paper aims to explore the impact of HNTs on plant biology and how sustainable plant adaptation can help mitigate the adverse effects of climate change.[3] Climate change has resulted in unprecedented changes in weather patterns, including increased frequency and intensity of heat waves. HNTs are among the most severe consequences of global warming and have a significant impact on plant biology. According to a study by Prasad et al.,[4] HNTs can cause reduced crop yields, delayed maturity, and decreased seed production in many plant species. Furthermore, these extreme weather conditions can lead to an increased susceptibility to pests and diseases, causing significant economic losses in agriculture.[5]

As climate change continues to impact global temperatures, sustainable adaptation strategies are crucial for plant survival. Various studies have explored the physiological and molecular mechanisms that plants utilize to cope with HNTs, such as antioxidant defense systems, heat shock proteins, and gene regulation.[6] In addition, several studies have proposed potential strategies to breed crops that are more resilient to these conditions, including selecting for heat tolerance traits and enhancing genetic diversity.[7] Climate change has become one of the most pressing issues of the modern era, with its potential to bring about significant changes in various ecological systems. One of the key consequences of climate change is the rise in temperature, particularly the HNT, which is anticipated to have a considerable impact on plant biology. According to Wang et al.,[8] HNT can result in changes in plant morphology, physiology, and metabolism, which can ultimately affect their growth and development. Similarly, Firon et al.[9] found that HNTs affect the expression of genes involved in various plant processes, including photosynthesis, respiration, and stress response.

On a global scale and at the farm level, minimum night-time temperatures are increasing significantly faster than maximum daytime temperatures, and this trend is projected to continue.[10,11] Higher minimum night temperatures have been proven to have a significant negative effect on rice yield in both field experiments[12] and controlled environment studies.[13] To produce diurnal temperature tolerance in rice, inbreeding for high day temperature tolerance in mega varieties of rice is progressing.[14] Therefore, we need to increase our efforts to counteract this worrying trend. A broad variety of entries must be evaluated for their reactivity to HNTs to find contrasting entries and better comprehend and research the physiological and molecular processes that generate tolerance. Reduced panicle density,[10] grain weight[13,15], and spikelet fertility[10] are all hypothesized to be to blame for the yield penalty under HNT.[16] Enhanced respiratory rates and membrane leaks help to explain these results to some extent. However, the bulk of the aforementioned conclusions were based on the efficiency of certain genotypes, such as IR72,[17] Cocodrie,[16] and Akita-63,[13] which were all investigated in lab settings. There was a significant void in our understanding of the physiological and biochemical responses of diverse rice genotypes to exposure to HNTs in real-world field conditions. Although it has not been thoroughly researched, it has been hypothesized that the temperature at night may affect the dynamics of blooming the following morning.[18] Rice's ability to successfully fill its grains is governed by photoassimilates produced either during grain filling (postanthesis) or transferred from the reserve pool of the vegetative tissues (preanthesis). Growing attention is being paid to the effects of temperature incidence on rice farming in light of projections that global average surface air temperatures may rise by 1.8–4.5°C by the turn of the century, with minimum night-time temperatures increasing at faster rates than maximum daytime ones.[19] According to Karl et al.,[20] there has been a rise in night-time temperatures that is roughly 3 times as big as the comparable increase in daytime temperatures. The 2010 summer was incredibly warm in Arkansas. In 2010, the average night-time air temperature (NTAT) measured during kernel development was approximately 5°C higher than in 2009. Six rice cultivars were field-grown at five different locations in Arkansas in 2010, and the average NTAT measured during kernel development was approximately 5°C higher than in 2009.[21] These elevated NTATs caused decreases in milling yield, amylose content, and total protein content; increases in milled rice whiteness, chalkiness, total lipid content, gelatinization temperature, and flour paste peak viscosity; and a certain degree of high-temperature (HT) incidence susceptibility in all cultivars tested.[22] Most of the quality changes caused by greater NTATs in field-grown rice were comparable to those seen in the greenhouse or temperature-controlled experiments.[23] Since the literature often associates high gelatinization temperature and enthalpy with high amylose concentration, the pattern of reduced amylose concentration and increased gelatinization temperature was uncommon among the effects of increasing NTAT. The ratio of amylose to amylopectin and the percentage distribution of amylopectin branch chains impact the functionality of starch. As the primary component of starch, amylopectin must be thoroughly investigated to appreciate the mechanisms behind the detrimental impacts of excessive NTAT on rice processing and end-use quality. The findings of studies done in environments with controlled temperatures regarding the effect of elevated temperature on the fine structure of amylopectin are not uniform. Lisle et al.[24] determined that the amylopectin chain length distribution of three cultivars grown at different daily mean temperatures did not vary. According to Suzuki et al.,[25] relative amounts of amylopectin long chains [degree of polymerization (DP > 25)] increased in all starches when maturation occurred at 28°C as compared to 21°C. According to Cheng et al.,[26] the fraction of amylopectin short chains (DP 22) varied in opposite ways that it increased in the low-amylose cultivar while it decreased in the high-amylose cultivar. According to Counce et al.,[27] the proportion of DP 13–24 chains in two long-grain cultivars increased by nearly one percentage point at a night-time temperature of 24°C compared to 18°C. Cooper et al.[23] found, however, that NTAT effects on amylopectin fine structure are cultivar-specific since hybrid and medium-grain cultivars were unaffected but the two long-grain cultivars used in their research did not react consistently to the temperature treatments. In more recent research, Wei et al.[28] discovered that an increase in temperature (32 vs

22°C) led to an increase in DP 10–13 and DP > 42 chains, whereas DP 5–9 and DP 15–22 chains decreased. Thus, understanding the impact of HNT on plant biology is crucial for developing sustainable adaptation strategies to mitigate the effects of climate change.

3.2 EFFECT OF HNTS ON RESPIRATION, MEMBRANE STABILITY, AND ANTIOXIDANT CAPACITY

By the end of the 21st century, global circulation models predict that all greenhouse gas concentrations will have increased, causing the global temperature to rise by 1.4–5.8°C. Temperatures at night are anticipated to rise higher than those during the day.[29] At the International Rice Research Institute (IRRI) in Manila, the Philippines, temperatures at night rose by 1.13°C over 25 years (1979 to 2003).[10] 10% less rice grain may be produced for every 1°C increase in night-time temperature.[10] It can therefore be advantageous for global rice production and food stability to discover and develop management strategies, such as the application of agrochemicals to prevent or counteract the detrimental impacts of high night-time temperatures.

Given that respiratory processes may consume a higher proportion of total photosynthates as a result of expected increases in plant respiration in response to climatic warming, this is a severe worry.[30] To better understand how the environment affects respiratory processes, respiration is frequently divided into the functional components of construction (growth), maintenance, and ion uptake.[31–32] According to Penning de Vries,[33] maintenance respiration is the most responsive to environmental changes and is primarily related to the turnover of proteins and lipids and the maintenance of ion concentration gradients across membranes. High nocturnal temperatures are viewed as undesirable because they can increase respiration, which can harm yield. In many plants, increased respiration rates brought on by HTs can result in the generation of reactive oxygen species (ROS).[34] Enzymic and nonenzymatic antioxidants reduce the damaging effects of ROS under normal physiological conditions; nevertheless, under stressful circumstances, oxidant levels might surpass the antioxidant levels, resulting in cell damage.[35] Increased ROS damage to cells can reduce membrane thermal stability (MTS), which will disturb the passage of water, ions, and organic solvent across plant membranes. This will affect the production, consumption, transport, and accumulation of carbon (Fig. 3.1).[36] Examining MTS, which detects electrolyte leakage from tissues subjected to stressors including dehydration, heat, and freezing is a standard technique for evaluating damage to membranes.[37] Under heat-stressed conditions, MTS was favorably correlated with wheat (*Triticum esculentum* L.) yield performance.[38] In some situations, a plant's antioxidant content is strongly related to how well it can withstand stress. The antioxidant state of the plant affects how much damage is caused by ROS. Plants pretreated with the putative preventative exogenous effectors (chemicals) utilized in this investigation, -tocopherol (vitamin E), glycine betaine (GB), or salicylic acid (SA), shown induced temperature tolerance and protection against oxidative damage.[39] While GB increases resistance to HTs by shielding certain enzymes (such as Rubisco and citrate synthase) against heat-induced inactivation, -tocopherol is one of the most effective single-oxygen quenchers and is a powerful antioxidant.[40] SA is crucial for protecting plants from oxidative damage by detoxifying superoxide radicals and stabilizing trimers of heat shock transcription factors. SA increases temperature tolerance and participates in calcium signaling.[41] Although antioxidants are important for stress tolerance, little is known about how rice thermal tolerance reacts to these preventative exogenous effects (vitamin E, GB, and SA).

3.3 PLANT GROWTH AND DEVELOPMENT UNDER HIGH NIGHT TEMPERATURES

It is generally known that HTs affect the growth and development of wheat and other crops. HTs decrease leaf photosynthetic rate, increase embryo abortion, damage photosynthetic membranes (thylakoids), and result in chlorophyll loss, lower grain number, and decreased grain fling duration and rates.[42,43] HTs negatively impact cell metabolism at the molecular level and alter the pattern of protein synthesis. In addition to inducing the synthesis and accumulation of numerous new proteins, supra-optimal temperatures also suppress the production of the normal complement of cellular proteins. These new proteins include heat shock proteins, Rubisco activase, chloroplast glyceraldehyde 3-phosphate dehydrogenase, and chloroplast protein synthesis elongation factor.[44,45]

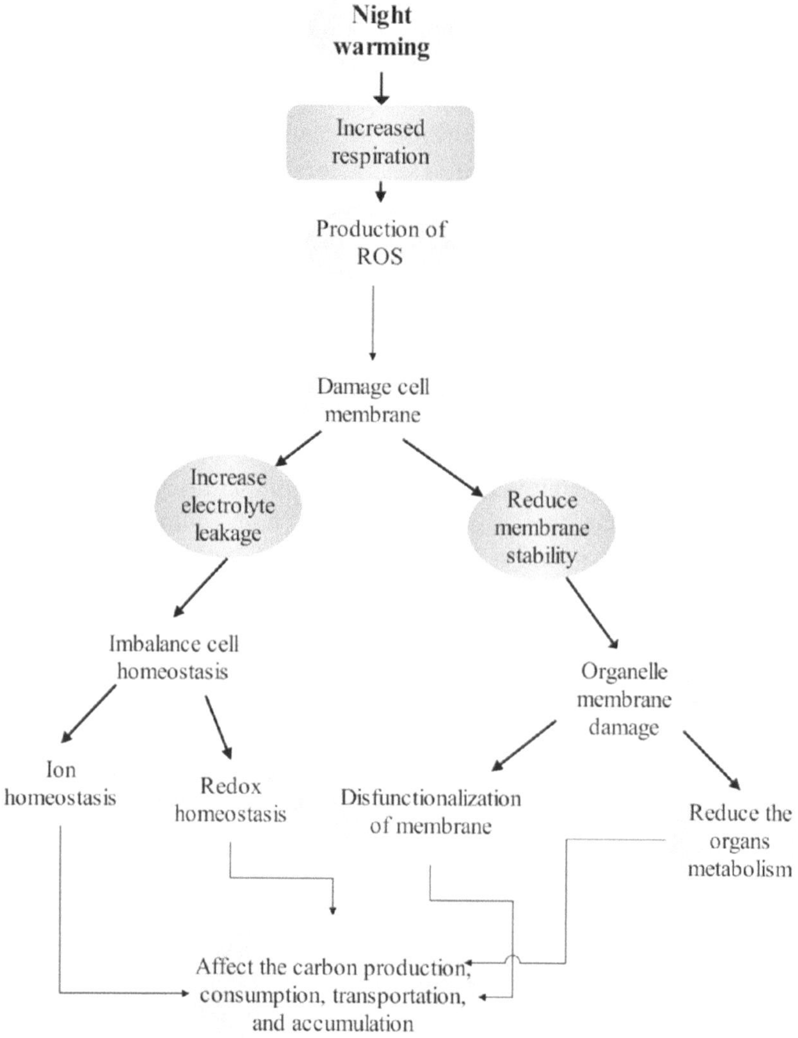

FIFURE 3.1 Effect of night warming on the cell membrane stability and homeostasis.

The majority of research on how hot weather affects agricultural plants has not made a distinction between daytime and nocturnal temperature regimes. The duration of critical periods, growth rates, and sensitivity of crops to minimum and maximum temperatures can all vary.

High overnight temperatures, for instance, are more detrimental to grain growth than high daytime temperatures in the case of rice.[15] Additionally, historical yields in the Yanqui Valley of Mexico were substantially connected with nightly minimum temperatures but not with daytime maximum temperatures, according to modeling research.[46] Because anthropogenic climate change is characterized by greater increases in night-time minimum temperatures than daytime maximum temperatures, it has been suggested that a better understanding of the plant responses to high night-time temperatures is necessary to better quantify and reduce uncertainties in assessments of the impact of climate change.[46] Very little is known about how variable overnight temperatures throughout the reproductive phases of wheat affect many physiological, growth, developmental, and yield processes (while maintaining a constant daylight temperature).

Night-time temperatures have both direct and indirect effects on plant physiology, morphology, growth, and yield. Photosynthesis is the plant's physiological activity most negatively affected by HNT and low night temperature (LNT).[47] HNT improved A_{net} for C_3 species, but it had a detrimental effect on C_4 species or no effect at all. Fluorescence parameters include photochemical efficiency of PSII (Fv/Fm), PSII quantum yield (PSII), excitation-to-emission ratio (ETR), stomatal conductance (g_s), and enzyme activities related to carbon fixation. HNT and plant height were revealed to positively correlate among diverse PFTs.[48] Understanding how plants react to changes in NTs may depend on a variety of factors, including the experimental design (such as treatment duration and growing facility) and species functional groupings. In a field experiment, the reduction of N and nonstructural material translocation after blooming was linked to a considerable decrease in rice production at HNT.[49] Reduced dry matter allocation to grain in a pot-growing experiment was a worry for yield loss because of lower spikelet fertility during the reproductive stage.[17] The length of the experiments affected how the A_{net} deduction from HNT and LNT varied.[50] However, the distinction between reactions to various treatment durations or to expanding facilities is unclear. Theoretically and practically, confirming the results of experimental methods is crucial for assisting agriculture in selecting the best cultivation strategy to reduce negative impacts brought on by HNT. The effect of night warming on vital plant processes is highlighted in Figure 3.2.

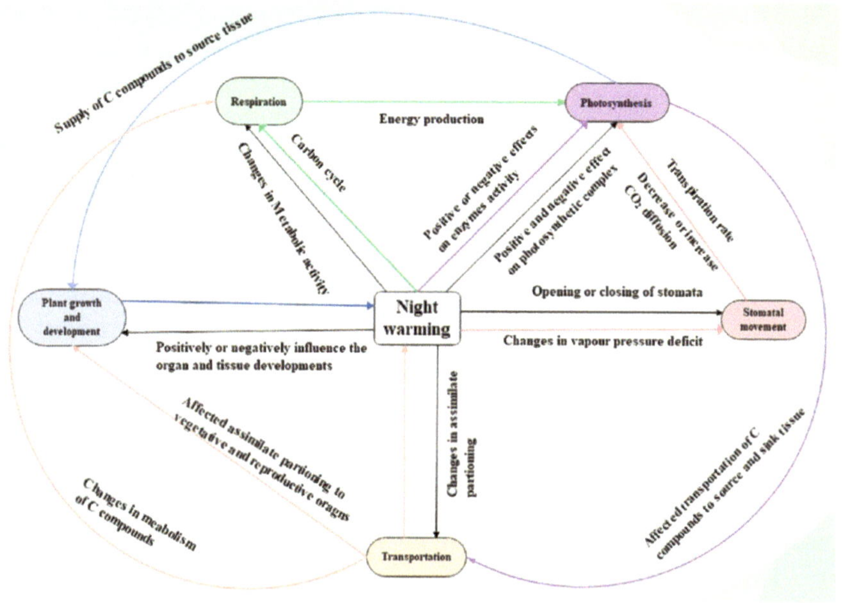

FIGURE 3.2 Diagrammatic representation of the impact of night warming on the plants vital processes.

3.4 HIGH NIGHT TEMPERATURE AND REPRODUCTIVE GROWTH

Lower rice yields would occur from a rise in NT caused by global warming.[10] The yield of wheat per acre is determined by the number of productive tillers per unit area, fertile spikelets per panicle, and grain weight—all of which are temperature sensitive.[51] HTs may cause sterility by impeding anther dehiscence, pollination, pollen germination on the stigma, pollen tube expansion, or the early stages of fertilization.[52] HTs not only shrink grain diameters but also reduce the amount of assimilation that spikelets get.[27] Damaged membranes brought on by heat-induced ROS and higher respiration rates may be the cause of the spikelet's decreased assimilate supply.[53]

High night-time temperatures, according to experiments, accelerated blooming time and reduced seed size and seed output. According to Egli and Wardlaw,[54] an HNT was linked to a smaller seed size decrease and a slower pace of seed development. On the contrary, it has been noted that a high daytime temperature has a greater impact on the lowering of seed output than an HNT.[55] It is still challenging to correctly predict the effects of night-time temperature on soybean reproductive development.

Increases in flower and pod production may help to offset the decline in seed output brought on by a warm night-time temperature. In many crops, not only is the reproductive process more vulnerable to heat stress than vegetative growth, but certain phases of the flower's development are also more vulnerable. HTs, exposure times, and length all dramatically decreased the number of pegs and pods per plant in groundnut. Meiosis in tomato mother cells, which occurs 8–9 days before anthesis and occurs in both mega- and microspore cells, is very susceptible to HTs.[56]

Additionally, according to Iwahori, ovules exposed to a temperature of 40°C for 18 h after pollination aborted, perhaps as a result of a reduction in pollen tube development and endosperm deterioration.[56] *Brassica napus* preanthesis HT treatments were also demonstrated to decrease pollen fertility.[57] The number of pollen grains released and germinated increased when tomato plants grown at 32/26°C were placed at 28/22°C for 10 days before anthesis, leading to more fruit set than when plants remained in the HT treatment (32/26°C).[58]

The environmental temperature during kernel development is a key factor in producing unanticipated changes in rice grain quality. HNTs, for instance, have an impact on kernel growth and dramatically lower head rice production. Similar effects were seen in grains subjected to HT stress, including an increase in the development of chalkiness, a drop in amylose content, and a reduction in grain size. In addition, HTs may alter the qualities of flour, bread, and other physicochemical elements of grain crops, such as variations in the protein content of flour during the grain filling phase. During the grain filling stage, HNT exposure altered the enzymatic activity, resulting in irregular starch granule packing that increased chalkiness. However, it remains unknown how HTs impact grain structure and cooking quality. HNT affected the amylose content, increased the chalkiness, and decreased head rice production and grain width in several genotypes under study. The performance of rice against rising temperatures may be enhanced by properly using the genotypic modifications in rice as well as by using various chemical applications and exact combinations. Studies on the possible impacts of these tactics on the qualitative features of rice are required owing to the paucity of information on these two processes.

Peng et al.[10] hypothesized that NTAT affected rice and other agricultural field production more than daytime temperature. The International Rice Research Institute (IRRI) detected a 1.13°C increase in night-time temperatures between 1992 and 2003. A later study on rice cultivars in controlled conditions showed that raising night temperatures from 21 to 32°C reduced yields, pollen germination, spikelet sterility, and respiration rates.[12] Cheng et al.[59] studied carbon and nitrogen absorption in rice plants in a controlled setting and found that increasing NTATs drastically increased night respiration, which may affect grain yields. Morita et al.[60] studied 22 and 34°C NTATs in a greenhouse. Higher NTAT levels lowered grain bulk, kernel growth, and endosperm cell size. Warm temperatures (typically above 22°C) interfere with rice's metabolic activities, affecting production and pre- and postharvest quality. Chalkiness, a critical quality concern in rice kernels, has environmental and genetic factors.[24] It is assumed to be caused by elevated temperatures throughout vital rice growth stages. Chalk is generated when uneven packing of starch granules (amyloplasts) causes air gaps in the endosperm of rice kernels, resulting in opaque white sections. According to Cheng et al.,[26] temperature-sensitive grain-bound starch synthase influences the grain-filling stage (GBSS). During grain-filling, GBSS activity was greater at 22 than 32°C. Increased chalk formation and reduced amylose content in rice may be due to decreased enzyme activity and physiological functioning rates after high-NTAT exposure.[23]

According to studies, warming affects the morpho-physiological activities of plants, changing crop development and growth, photosystem II (PSII) functioning, antioxidant systems, and grain filling time. Wheat grain weight rises by 1.1°C at night because both better and inferior grains load their grains faster.[61] Wheat's grain-filling phase is characterized by a close relationship between yield and net photosynthetic assimilation. The final grain dry weight of winter wheat is mostly determined by the photosynthesis of the flag leaf. It is crucial to examine the reaction of the flag leaf to winter and springtime temperature increases. Photosynthesis is the most temperature-sensitive activity in plants. Changes in the thermal optimum and increases in photosynthetic rates at new growth temperatures indicate that some species may adapt to temperature shifts.[61]

3.5 HIGH NIGHT TEMPERATURE ON YIELD AND QUALITY ATTRIBUTES

Many of the high-yielding semi-dwarf rice cultivars available today are sensitive to heat, with yields declining proportionately to daily temperatures exceeding 33°C. The major reasons why seed set and output decline with higher day temperatures include increased floral sterility decreased pollen grain viability during anthesis, and, to a lesser extent, limited carbon transport to the developing grains. Many rice cultivars develop higher than 50% infertility when daytime temperatures surpass 36°C.[62]

It is generally recognized that rising daytime temperatures have a negative influence on rice seed sets, but less is known about the effects of HNTs when compared to stable daytime temperatures. Recent studies suggest that the anticipated global warming may be uneven, with night-time temperature increases being substantially greater than daylight temperature increases. In sunny, outdoor environment chambers, the effects of increasing night temperatures at two constant day temperatures—29/21, 29/25, 29/29°C and 33/25, 33/29, 33/33°C—as well as increasing day/night temperatures—29/21, 33/25, and 36129°C—on rice growth and production were studied (cv. IR 72). At harvest, both plant biomass and grain production were significantly reduced when daytime and night-time temperatures were raised to 36/29°C. In contrast, doing so at a constant daytime temperature of 33°C (8 h) had a significant negative influence on grain production, mostly because of reduced seed set. Increasing night-time temperatures at a constant daily temperature of 29°C had no obvious effect on growth or yield. When night-time temperatures rose to 33°C, grain output decreased similarly to when daytime temperatures were 36 and 29°C. The results of this experiment indicate that increased night-time temperatures by themselves may make rice more susceptible to sterility which would reduce seed set and grain output.[63] The effect of HNTs on grain quality traits was highlighted in Table 3.1.

An aspect of this warming impact that is somewhat less well-known is yield reductions that are especially linked to rises in minimum (i.e., night-time) temperatures. The fact that global night-time temperature values are rising at a rate that is 1.4 times faster than daytime temperature values has the effect of making warming trends more asymmetrical. The subsequent nocturnal warming events may cover a wider geographic area than daylight warming and last for a longer period throughout the crop cycle. As a consequence, while originally somewhat contentious, a number of follow-up studies ultimately confirmed the detrimental effects of increasing night-time temperatures on agricultural yields for a variety of crops, including tomato, and rice.[64,65] Winter wheat cultivars that are regularly grown in the US Great Plains showed yield declines of an average of 6% for every 1°C increase in night-time temperatures during the seed-fill phase in experiments combining field-based infrastructure with cyber-physical

TABLE 3.1 Effect of HNTs on Quality Attributes of Some Important Crops.

Crop	Treatments	Effect of high night temperature	References
Oryza sativa L.	HDT: DT; 35 ± 2°C; NT; 28 ± 2°C HNT: DT; 28 ± 2°C; NT; 32 ± 2°C	Reduce 6–21 and 7–27% grain appearance and grain milling attributes respectively, whereas it increased by 76–134% endosperm chalkiness	[68]
O. sativa L.	NT: 24, 30, and 35°C	Declined amylose content, and gel consistency with increased NT	[69]
O. sativa L.	NT; 26.7°C	Compact arrangement of endosperms, smaller and imperfect grains, decreased viscosity, and increased chalkiness	[70]
Glycine max (L.) Merr.	DT/NT of 30/20, 30/30, 35/20, and 35/30°C	Decreased Ca and Mn, and alters mineral nutrients composition in seed	[71]
O. sativa L.	Control: 16 h 30 ± 1°C: 8 h 23 ± 1°C, light : dark HNT: 16 h 30 ± 1°C : 8 h 28 ± 1°C, light : dark	Affect grain length and width	[72]
Triticum aestivum L.	DT: 28°C HNT: 15,18, 21, 23, 25, and 27°C	Reduced grain length and width 6 and 10% and 18 and 21%, respectively, at 25 and 27°C	[73]
T. aestivum L.	Increased NT by 1.8°C	Increased ratio of glutenin to gliadin, and smooth and dense gluten	[74]
O. sativa L.	Control: 24°C HNT: 29°C	Higher chalkiness and waxy grains	[75]
Hordeum vulgare L.	Increased NT by 4°C	Reduced grain weight	[76]
O. sativa L.	Control: 24°C HNT: 29°C	Reduce grain width and decrease amylose content	[77]

*HDT; high day temperature, HNT; high night temperature, DT; day temperature; NT; night temperature

systems.[66] Field-grown spring wheat and rice that received treatments throughout the flowering and seed-filling stages of growth observed yield decreases of 4–7% for every 1°C increase in night-time temperatures.[67] We still know very little about the physiological underpinnings of night-time warming, which is threatening global food security by causing yield losses everywhere. Furthermore, it is often believed that these decreases are primarily caused by temperature rises at night.

3.6 CONCLUSIONS

Over the last 10 years, there has been a lot of research done on how asymmetric warming affects plant species negatively. The screening of several cultivars to find HNT-tolerant variants for various grain species is a big issue. To define natural variation, great care is made to enhance the repeatable experimental setup of HNT settings. Cereals were cultivated using overhead infrared heaters in greenhouse testing, in situ temperature-controlled field chambers, and field tents with heaters and ventilators. Recent studies detail heat tolerance research procedures and compare measuring HNT impacts on grain yield in controlled environmental chambers to field experiments. Alternatives are necessary for the development of HNT-tolerant cultivars since crops react differently to HNT and the aforementioned HNT research facilities are difficult to establish in all stress-prone locales. We think that improving omics approaches is critical for producing molecular markers suitable for marker-assisted breeding or molecular engineering of HNT-tolerant crops. This approach will aid current efforts to develop HNT-tolerant cultivars using conventional and molecular breeding methods.

KEYWORDS

- **climate change**
- **high night temperatures**
- **plant biology**
- **photosynthesis**
- **respiration**
- **growth**
- **yield**
- **molecular mechanisms**
- **biochemical mechanisms**
- **adaptation**
- **sustainability**

REFERENCES

1. Shabbir, R.; Singhal, R. K.; Mishra, U. N.; Chauhan, J.; Javed, T.; Hussain, S.; Kumar, S.; Anuragi, H.; Lal, D.; Chen, P. Combined Abiotic Stresses: Challenges and Potential for Crop Improvement. *Agronomy* **2022,** *12* (11), p.2795.
2. Dey, P.; Datta, D.; Pattnaik, D.; Dash, D.; Saha, D.; Panda, D.; Bhatta, B. B.; Parida, S.; Mishra, U. N.; Chauhan, J.; Pandey, H. Physiological, Biochemical, and Molecular Adaptation Mechanisms of Photosynthesis and Respiration Under Challenging Environments. In *Plant Perspectives to Global Climate Changes*; Academic Press, 2022; pp 79–100.
3. Jing, P.; Wang, D.; Zhu, C.; Chen, J. Plant Physiological, Morphological and Yield-Related Responses to Night Temperature Changes Across Different Species and Plant Functional Types. *Front. Plant Sci.* **2016,** *7*, 1774.
4. Prasad, P. V.; Bheemanahalli, R.; Jagadish, S. K. Field Crops and the Fear of Heat Stress—Opportunities, Challenges and Future Directions. *Field Crops Res.* **2017,** *200*, 114–121.
5. Das, D. K.; Singh, J.; Vennila, S. Emerging Crop Pest Scenario Under the Impact of Climate Change—A Brief Review. *J. Agric. Phy.* **2011,** *11*, 13–20.
6. Shafqat, W.; Jaskani, M. J.; Maqbool, R.; Chattha, W. S.; Ali, Z.; Naqvi, S. A.; Haider, M. S.; Khan, I. A.; Vincent, C. I. Heat Shock Protein and Aquaporin Expression Enhance Water Conserving Behavior of Citrus Under Water Deficits and High Temperature Conditions. *Environ. Exp. Bot.* **2021,** *181*, 104270.
7. Hatfield, J. L.; Boote, K. J.; Kimball, B. A.; Ziska, L. H.; Izaurralde, R. C.; Ort, D.; Thomson, A. M.; Wolfe, D. Climate Impacts on Agriculture: Implications for Crop Production. *Agron. J.* **2011,** *103* (2), 351–370.
8. Wang, J.; Xi, Z.; He, X.; Chen, S.; Rossi, S.; Smith, N. G.; Liu, J.; Chen, L. Contrasting Temporal Variations in Responses of Leaf Unfolding to Daytime and Night Time Warming. *Glob. Chang. Biol.* **2021,** *27* (20), 5084–5093.
9. Firon, N.; Shaked, R.; Peet, M. M.; Pharr, D. M.; Zamski, E.; Rosenfeld, K.; Althan, L.; Pressman, E. Pollen Grains of Heat Tolerant Tomato Cultivars Retain Higher Carbohydrate Concentration Under Heat Stress Conditions. *Sci. Hortic.* **2006,** *109* (3), 212–217.
10. Peng, S.; Huang, J.; Sheehy, J. E.; Laza, R. C.; Visperas, R. M.; Zhong, X.; Centeno, G. S.; Khush, G. S.; Cassman, K. G. Rice Yields Decline with Higher Night Temperature from Global Warming. *Proc. Natl. Acad. Sci.* **2004,** *101* (27), 9971–9975.
11. Gregory, J.; Stouffer, R. J.; Molina, M.; Chidthaisong, A.; Solomon, S.; Raga, G.; Friedlingstein, P.; Bindoff, N. L.; Le Treut, H.; Rusticucci, M.; Lohmann, U. Climate Change: The Physical Science Basis, 2007.
12. Nagarajan, S.; Jagadish, S. V. K.; Prasad, A. H.; Thomar, A. K.; Anand, A.; Pal, M.; Agarwal, P. K. Local Climate Affects Growth, Yield and Grain Quality of Aromatic and Non-Aromatic Rice in Northwestern India. *Agric. Ecosys. Environ.* **2010,** *138* (3–4), 274–281.
13. Kanno, K.; Makino, A. Increased Grain Yield and Biomass Allocation in Rice Under Cool Night Temperature. *Soil Sci. Plant Nut.* **2010,** *56* (3), 412–417.

14. Zhi-Hong, Z.; Li, S.; Wei, L.; Wei, C.; Ying-Guo, Z. A Major QTL Conferring Cold Tolerance at the Early Seedling Stage Using Recombinant Inbred Lines of Rice (*Oryza sativa* L.). *Plant Sci.* **2005,** *168* (2), 527–534.
15. Morita, S.; Shiratsuchi, H.; Takahashi, J.; Fujita, K. Effect of High Temperature on Grain Ripening in Rice Plants: Analysis of the Effects of High Night and High Day Temperatures Applied to the Panicle and Other Parts of the Plant. *Jpn J. Crop Sci.*, **2004**.
16. Mohammed, A. R.; Tarpley, L. Impact of High Night time Temperature on Respiration, Membrane Stability, Antioxidant Capacity, and Yield of Rice Plants. *Crop Sci.* **2009,** *49* (1), 313–322.
17. Cheng, W.; Sakai, H.; Yagi, K.; Hasegawa, T. Interactions of Elevated [CO_2] and Night Temperature on Rice Growth and Yield. *Agric. Forest Meteor.* **2009,** *149* (1), 51–58.
18. Kobayashi, K.; Matsui, T.; Murata, Y.; Yamamoto, M.; Percentage of Dehisced Thecae and Length of Dehiscence Control Pollination Stability of Rice Cultivarsat High Temperatures. *Plant Prodn Sci.* **2011,** *14* (2), 89–95.
19. Bernstein, L.; Bosch, P.; Canziani, O.; Chen, Z.; Christ, R.; Riahi, K. IPCC, 2007: Climate Change: Synthesis Report, 2007.
20. Karl, T. R.; Kukla, G.; Razuvayev, V. N.; Changery, M. J.; Quayle, R. G.; Heim Jr, R. R.; Easterling, D. R.; Fu, C. B. Global Warming: Evidence for Asymmetric Diurnal Temperature Change. *Geophy. Res. Lett.* **1991,** *18* (12), 2253–2256.
21. Lanning, S. B.; Siebenmorgen, T. J.; Ambardekar, A. A.; Counce, P. A.; Bryant, R. J. Effects of Night Time Air Temperature During Kernel Development of Field-Grown Rice on Physicochemical and Functional Properties. *Cereal Chem.* **2012,** *89* (3), 168–175.
22. Lanning, S.; Siebenmorgen, T.; Effects of Preharvest Night Time Air Temperatures on Whiteness of Head Rice. *Cereal Chem.* **2013,** *90* (3), 218–222.
23. Cooper, N. T. W.; Siebenmorgen, T. J.; Counce, P. A. Effects of Night-time Temperature During Kernel Development on Rice Physicochemical Properties. *Cereal Chem.* **2008,** *85* (3), 276–282.
24. Lisle, A. J.; Martin, M.; Fitzgerald, M. A. Chalky and Translucent Rice Grains Differ in Starch Composition and Structure and Cooking Properties. *Cereal Chem.* **2000,** *77* (5), 627–632.
25. Suzuki, Y.; Sano, Y.; Ishikawa, T.; Sasaki, T.; Matsukura, U.; Hirano, H. Y. Starch Characteristics of the Rice Mutant du2-2 Taichung 65 Highly Affected by Environmental Temperatures During Seed Development. *Cereal Chem.* **2003,** *80* (2), 184–187.
26. Cheng, F.; Zhong, L.; Zhao, N.; Liu, Y.; Zhang, G. Temperature Induced Changes in the Starch Components and Biosynthetic Enzymes of Two Rice Varieties. *Plant Growth Regul.* **2005,** *46*, 87–95.
27. Counce, P. A.; Bryant, R. J.; Bergman, C. J.; Bautista, R. C.; Wang, Y. J.; Siebenmorgen, T. J.; Moldenhauer, K. A. K.; Meullenet, J. F. Rice Milling Quality, Grain Dimensions, and Starch Branching as Affected by High Night Temperatures. *Cereal Chem.* **2005,** *82* (6), 645–648.
28. Wei, K. S.; Yang, W. L.; Jilani, G.; Zhou, W. J.; Liu, G. K.; Chaudhry, A. N.; Cao, Z. Z.; Cheng, F. M. Effect of High Temperature on the Enzymatic Activities and

Transcriptional Expression of Starch Debranching Enzyme (DBE) Multiple Isoforms in Developing Rice Endosperms. *J. Anim. Plant Sci.* **2012,** *22*, 97–107.
29. Alward, R. D.; Detling, J. K.; Milchunas, D. G. Grassland Vegetation Changes and Nocturnal Global Warming. *Science* **1999,** *283* (5399), 229–231.
30. Paembonan, S. A.; Hagihara, A.; Hozumi, K. Long-Term Respiration in Relation to Growth and Maintenance Processes of the Aboveground Parts of a Hinoki Forest Tree. *Tree Physiol.* **1992,** *10* (1), 101–110.
31. Lambers, H. Respiration in Intact Plants and Tissues: Its Regulation and Dependence on Environmental Factors, Metabolism and Invaded Organisms. *Higher Plant Cell Respiration* **1985**, 418–473.
32. Amthor, J. S. Evolution and Applicability of a Whole Plant Respiration Model. *J. Theoret. Biol.* **1986,** *122* (4), 473–490.
33. De Vries, F. P. The Cost of Maintenance Processes in Plant Cells. *Ann. Bot.* **1975,** *39* (1), 77–92.
34. McDonald, A. E.; Vanlerberghe, G. C. Alternative Oxidase and Plastoquinol Terminal Oxidase in Marine Prokaryotes of the Sargasso Sea. *Gene* **2005,** *349*, 15–24.
35. Kreiner, M.; Harvey, L. M.; McNeil, B. Oxidative Stress Response of a Recombinant *Aspergillus niger* to Exogenous Menadione and H_2O_2 Addition. *Enzyme Microb. Technol.* **2002,** *30* (3), 346–353.
36. Christiansen, M. N. The Physiology of Plant Tolerance to Temperature Extremes. *Crop Tolerance Suboptimal Land Conditions* **1978,** *32*, 173–191.
37. Sewelam, N.; Kazan, K.; Schenk, P. M. Global Plant Stress Signaling: Reactive Oxygen Species at the Cross-Road. *Front. Plant Sci.* **2016,** *7*, 187.
38. Reynolds, M. P.; Balota, M.; Delgado, M. I. B.; Amani, I.; Fischer, R. A. Physiological and Morphological Traits Associated with Spring Wheat Yield Under Hot, Irrigated Conditions. *Funct. Plant Biol.* **1994,** *21* (6), 717–730.
39. Hasanuzzaman, M.; Nahar, K.; Fujita, M. Extreme Temperature Responses, Oxidative Stress and Antioxidant Defense in Plants. *Abiotic Stress-Plant Responses and Applications in Agriculture* **2013,** *13*, 169–205.
40. Makela, P.; Karkkainen, J.; Somersalo, S. J. B. P. Effect of Glycinebetaine on Chloroplast Ultrastructure, Chlorophyll and Protein Content, and RuBPCO Activities in Tomato Grown Under Drought or Salinity. *Biologia Plantar.* **2000,** *43* (3), 471–476.
41. Larkindale, J.; Knight, M. R. Protection Against Heat Stress-Induced Oxidative Damage in Arabidopsis Involves Calcium, Abscisic Acid, Ethylene, and Salicylic Acid. *Plant Physiol.* **2002,** *128* (2), 682–695.
42. Morales, F.; Ancín, M.; Fakhet, D.; González-Torralba, J.; Gámez, A. L.; Seminario, A.; Soba, D.; Ben Mariem, S.; Garriga, M.; Aranjuelo, I. Photosynthetic Metabolism Under Stressful Growth Conditions as a Bases for Crop Breeding and Yield Improvement. *Plants* **2020,** *9* (1), p.88.
43. Fahad, S.; Adnan, M.; Hassan, S.; Saud, S.; Hussain, S.; Wu, C.; Wang, D.; Hakeem, K. R.; Alharby, H. F.; Turan, V.; Khan, M. A. Rice Responses and Tolerance to High Temperature. In: *Advances in Rice Research for Abiotic Stress Tolerance*; Woodhead Publishing, 2019; pp201–224.
44. Jagadish, S. K.; Way, D. A.; Sharkey, T. D.; Plant heat stress: Concepts directing future research. *Plant, Cell & Environ.* **2021,** *44* (7),1992–2005.

45. Lal, M. K.; Tiwari, R. K.; Gahlaut, V.; Mangal, V.; Kumar, A.; Singh, M. P.; Paul, V.; Kumar, S.; Singh, B.; Zinta, G. Physiological and Molecular Insights on Wheat Responses to Heat Stress. *Plant Cell Rep.* **2021**, 1–18.
46. Lobell, D. B.; Ortiz-Monasterio, J. I. Impacts of Day Versus Night Temperatures on Spring Wheat Yields: A Comparison of Empirical and CERES Model Predictions in Three Locations. *Agron. J.* **2007,** *99* (2), 469–477.
47. Prasad, P. V.; Djanaguiraman, M.; High Night Temperature Decreases Leaf Photosynthesis and Pollen Function in Grain Sorghum. *Funct. Plant Biol.* **2011,** *38* (12), 993–1003.
48. Lucidos, J. G.; Ryu, K. B.; Younis, A.; Kim, C. K.; Hwang, Y. J.; Son, B. G.; Lim, K. B. Different Day and Night Temperature Responses in *Lilium hansonii* in Relation to Growth and Flower Development. *Hortic. Environ. Biotechnol.* **2013,** *54*, 405–411.
49. Shi, W.; Muthurajan, R.; Rahman, H.; Selvam, J.; Peng, S.; Zou, Y.; Jagadish, K. S. Source–Sink Dynamics and Proteomic Reprogramming Under Elevated Night Temperature and Their Impact on Rice Yield and Grain Quality. *New Phytol.* **2013,** *197* (3), 825–837.
50. Prasad, P. V.; Djanaguiraman, M. High Night Temperature Decreases Leaf Photosynthesis and Pollen Function in Grain Sorghum. *Funct. Plant Biol.* **2011,** *38* (12), 993–1003.
51. Prasad, P. V.; Pisipati, S. R.; Ristic, Z.; Bukovnik, U. R. S. K. A.; Fritz, A. K. Impact of Night-time Temperature on Physiology and Growth of Spring Wheat. *Crop Sci.* **2008,** *48* (6), 2372–2380.
52. Wassmann, R.; Jagadish, S. V. K.; Heuer, S.; Ismail, A.; Redona, E.; Serraj, R.; Singh, R. K.; Howell, G.; Pathak, H.; Sumfleth, K. Climate Change Affecting Rice Production: The Physiological and Agronomic Basis for Possible Adaptation Strategies. *Adv. Agron.* **2009,** *101*, 59–122.
53. Frantz, J. M.; Cometti, N. N.; Bugbee, B. Night Temperature Has a Minimal Effect on Respiration and Growth in Rapidly Growing Plants. *Ann. Bot.* **2004,** *94* (1), 155–166.
54. Egli, D. B.; Wardlaw, I. F. Temperature Response of Seed Growth Characteristics of Soybeans 1. *Agron. J.* **1980,** *72* (3), 560–564.
55. Gibson, L. R.; Mullen, R. E. Influence of Day and Night Temperature on Soybean Seed Yield. *Crop Sci.* **1996,** *36* (1), 98–104.
56. Iwahori, S. High Temperature Injuries in Tomato. V Fertilization and Development of Embryo with Special Reference to the Abnormalities Caused by High Temperature. *J. Japanese Soc. Hortic. Sci.* **1966,** *35* (4), 379–386.
57. Morrison, M. J.; Stewart, D. W. Heat Stress During Flowering in Summer Brassica. *Crop Sci.* **2002,** *42* (3), 797–803.
58. Sato, S.; Peet, M. M.; Thomas, J. F. Physiological Factors Limit Fruit Set of Tomato (*Lycopersicon esculentum* Mill.) Under Chronic, Mild Heat Stress. *Plant, Cell Environ.* **2000,** *23* (7), 719–726.
59. Cheng, W.; Sakai, H.; Yagi, K.; Hasegawa, T. Combined Effects of Elevated [CO_2] and High Night Temperature on Carbon Assimilation, Nitrogen Absorption, and the Allocations of C and N by rice (*Oryza sativa* L.). *Agric. Forest Meteor.* **2010,** *150* (9), 1174–1181.

60. Morita, S.; Yonemaru, J. I.; Takanashi, J. I. Grain Growth and Endosperm Cell Size Under High Night Temperatures in Rice (*Oryza sativa* L.). *Ann. Bot.* **2005,** *95* (4), 695–701.
61. Chen, J.; Tian, Y.; Zhang, X.; Zheng, C.; Song, Z.; Deng, A.; Zhang, W. Night-time Warming Will Increase Winter Wheat Yield Through Improving Plant Development and Grain Growth in North China. *J. Plant Growth Regul.* **2014,** *33*, 397–407.
62. Khan, A. H.; Min, L.; Ma, Y.; Zeeshan, M.; Jin, S.; Zhang, X.; High-Temperature Stress in Crops: Male Sterility, Yield Loss and Potential Remedy Approaches. *Plant Biotechnol. J.* **2022.**
63. Mohammed, A. R.; Tarpley, L. Effects of High Night Temperature and Spikelet Position on Yield-Related Parameters of Rice (*Oryza sativa* L.) Plants. *Europ. J. Agron.* **2010,** *33* (2), 117–123.
64. Willits, D. H.; Peet, M. M. The Effect of Night Temperature on Greenhouse Grown Tomato Yields in Warm Climates. *Agri. Forest Meteor.* **1998,** *92* (3), 191–202.
65. Impa, S. M.; Raju, B.; Hein, N. T.; Sandhu, J.; Prasad, P. V.; Walia, H.; Jagadish, S. K. High Night Temperature Effects on Wheat and Rice: Current Status and Way Forward. *Plant, Cell Environ.* **2021,** *44* (7), 2049–2065.
66. Fang, S.; Tan, K.; Ren, S. Winter Wheat Yields Decline with Spring Higher Night Temperature by Controlled Experiments. *Sci. Agric. Sinica* **2010,** *43* (15), 3251–3258.
67. Mamrutha, H. M.; Rinki, K.; Venkatesh, K.; Gopalareddy, K.; Khan, H.; Mishra, C. N.; Kumar, S.; Kumar, Y.; Singh, G.; Singh, G. P. Impact of High Night Temperature Stress on Different Growth Stages of Wheat. *Plant Physiol. Rep.* **2020,** *25*, 707–715.
68. Fahad, S.; Hussain, S.; Saud, S.; Hassan, S.; Chauhan, B. S.; Khan, F.; Ihsan, M. Z.; Ullah, A.; Wu, C.; Bajwa, A. A.; Alharby, H.; Responses of Rapid Viscoanalyzer Profile and Other Rice Grain Qualities to Exogenously Applied Plant Growth Regulators Under High Day and High Night Temperatures. *PLoS One* **2016,** *11* (7), p.e0159590.
69. Coast, O.; Ellis, R. H.; Murdoch, A. J.; Quiñones, C.; Jagadish, K. S.; High Night Temperature Induces Contrasting Responses for Spikelet Fertility, Spikelet Tissue Temperature, Flowering Characteristics and Grain Quality in Rice. *Funct. Plant Biol.* **2014,** *42* (2), 149–161.
70. Song, X.; Du, Y.; Song, X.; Zhao, Q. Effect of High Night Temperature During Grain Filling on Amyloplast Development and Grain Quality in Japonica Rice. *Cereal Chem.* **2013,** *90* (2), 114–119.
71. Gibson, L. R.; Mullen, R. E. Mineral Concentrations in Soybean Seed Produced Under High Day and Night Temperature. *Canadian J. Plant Sci.* **2001,** *81* (4), 595–600.
72. Dhatt, B. K.; Paul, P.; Sandhu, J.; Hussain, W.; Irvin, L.; Zhu, F.; Adviento-Borbe, M. A.; Lorence, A.; Staswick, P.; Yu, H.; Morota, G. Allelic Variation in Rice Fertilization Independent Endosperm 1 Contributes to Grain Width Under High Night Temperature Stress. *New Phytol.* **2021,** *229* (1), 335–350.
73. Impa, S. M.; Vennapusa, A. R.; Bheemanahalli, R.; Sabela, D.; Boyle, D.; Walia, H.; Jagadish, S. K. High Night Temperature Induced Changes in Grain Starch Metabolism Alters Starch, Protein, and Lipid Accumulation in Winter Wheat. *Plant Cell Environ.* **2020,** *43* (2), 431–447.
74. Bahuguna, R. N.; Solis, C. A.; Shi, W.; Jagadish, K. S. Post-Flowering Night Respiration and Altered Sink Activity Account for High Night Temperature-Induced

Grain Yield and Quality Loss in Rice (*Oryza sativa* L.). *Physiologia Plantarum* **2017,** *159* (1), 59–73.

75. García, G. A.; Serrago, R. A.; Dreccer, M. F.; Miralles, D. J. Post-Anthesis Warm Nights Reduce Grain Weight in Field-Grown Wheat and Barley. *Field Crops Res.* **2016,** *195*, 50–59.

PART II

NIGHT WARMING EFFECT ON CRUCIAL PROCESSES AND METABOLISM

CHAPTER 4

Night Warming Responses on Crucial Plant Processes and Growth Stages: An Overview

SUDARSHANA RANJAN[1], MEGHA PANWAR[1], ANJALI[1], UDIT NANDAN MISHRA[2], PRAJJAL DEY[2], S.K. MEENA[3], DEBANJANA SAHA[4], AMBIKA PRASAD MISHRA[2], GURDEEP BAINS[1], SUDHIR KUMAR GURU[1], and RAJESH KUMAR SINGHAL[3], K. V. KRISHNAMURTHY,[1] T. PULLAIAH,[2] and BIR BAHADUR

[1]*Department of Plant Physiology, G.B. Pant University of Agriculture and Technology, Pantnagar, Uttarakhand, India*

[2]*Faculty of Agriculture, Sri Sri University, Cuttack, Odisha, India*

[3]*ICAR–Indian Grassland and Forage Research Institute, Jhansi, Uttar Pradesh, India*

[4]*Department of Biotechnology, Centurion University of Technology and Management, Jatni, Bhubaneswar, India*

ABSTRACT

Global climate change is one of the major threats of the modern era, influencing the global ecosystem. In global climate change, human development and anthropogenic activities are the major contributing factors and are expected to have more serious effects shortly. Greenhouse gas (GHG) emissions, elevated temperature and CO_2, drought, flood, nutrient toxicity, and deficiency are the major issues associated with global climate change

Impact of High Night Temperature on Plant Biology: Toward Sustainable Plant Adaptation to Climate Change. Rajesh Kumar Singhal, Raju Bheemanahalli, Saurabh Pandey, and Pratibha M. D. (Eds.)

and affecting the social and economical aspects of livelihood. The agricultural systems are also severely affected by global climate change and are expected to have huge yield penalties for major crops. Therefore, the prime objective of agriculturists is to develop a solution for future climate change and reduce the yield penalty. In recent times, not only global warming but also night warming is one of the most drastic factors in crops and is linked to changes in the physiological, biochemical, and molecular processes of plants. Therefore, this chapter deliberately discusses the global climate change effects on ecological systems and the effect of night warming on plant growth stages and crucial processes.

4.1 INTRODUCTION

Global climate change refers to the long-term changes in the earth's climate system due to anthropogenic activities, including fossil fuel burning, leading to the trapping of greenhouse gases (GHGs) in the atmosphere. It is a global issue that affects numerous ecological, environmental, sociopolitical, and socioeconomic components.[1,2] In the last few decades, several countries have experienced higher temperatures due to variability in climate change.[3–5] With the advent of the industrial revolution, the problem of global climate change grew exponentially. The environment is constantly changing due to human activities, and the challenge today is to predict how agriculture will respond to the changing climatic environment. Global climate change, which affects agriculture, is (i) changes in land use and land cover; (ii) a worldwide decline in biodiversity; (iii) changes in atmospheric gases; and (iv) climate changes.

One of the major driving forces for global climate change is the rapid increase in concentrations of GHGs in the atmosphere. Human activities have led to the buildup of GHGs, including water vapors, CO_2, methane, nitrous oxide, and chlorofluorocarbons (CFCs), resulting in global climate change. All these gases except water vapor are produced directly or indirectly by human activities such as burning fossil fuels, cutting forests and grasslands for land clearing, using CFCs in refrigerants, industrial processes, and aerosol sprays (Fig. 4.1). Natural hazards resulting from global climate change cause the incidence of stress like drought, submergence, and temperature stress, concerning which the plants must undergo severe physiological and metabolic alterations, including shuffling in primary and secondary metabolite production and their organelle distribution. These adjustments affect the plant's capacity to meet the expected yield attributes. Also, the night warming affects plant growth and development from seed germination to the maturity phase. Also, the physiological, biochemical, and molecular processes are affected differentially depending on the crop and stage of crop growth. Food production frameworks are truly defenseless against environmental variation, such as variation in precipitation and temperature, which causes a plethora of biotic and abiotic stress, thereby reducing national food security by influencing crop harvest. The variety in the rate and dissemination of crop pathogens causes backhanded impacts. A dry winter climate that coincides with sudden temperature and precipitation insurgences, albeit marginally with the generally dry season, may cause pathogen and pest resurgences.[13] Therefore, this chapter discussed the status of GHGs in the environment, the consequences of global climate change, and the effect of night warming on different processes and different plant growth stages.

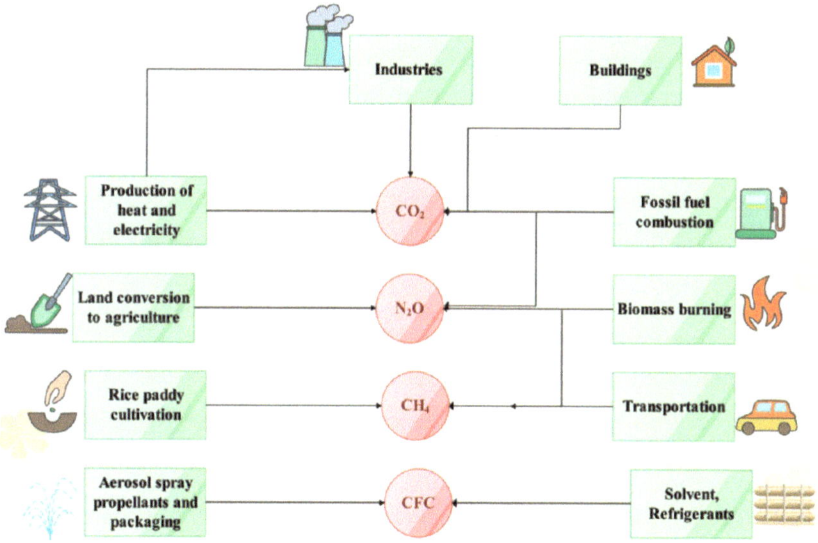

FIGURE 4.1 Greenhouse gases and their anthropogenic sources.

4.2 CURRENT AND FUTURE STATUS OF GLOBAL WARMING

The entire history of scientific discoveries and developments on climate change can be divided into seven periods:

1. Early and mid-1800, when changes in paleoclimate were first suspected and the natural greenhouse effect was identified.
2. Late 1800s, when scientists first argued that human emissions of GHGs induced climate change.
3. Early 1900 to mid-1900's period of controversy and disinterest in climate change.
4. 1950–1960 period of concern and increasing urgency.
5. 1970s: a period when scientists increasingly predicted global warming.
6. 1980–1988—period of scientific consensus about climate change.
7. 1988–Present modern period.

Considerable efforts have been expended to analyze the current historical climate record during the 20th century to establish whether any trends could be attributed to climate change. GHGs are the major reason, as they are found to remain in the atmosphere for extended durations.[6–8] Even if emissions stopped rising, atmospheric concentrations of GHGs would continue to rise and remain elevated for hundreds of years. Moreover, surface air temperatures would continue to rise even if concentrations were stabilized and the composition of the existing atmosphere stayed unchanged, which would require a drastic reduction in current GHG emissions. This is because it takes decades for the oceans, which store heat, to fully respond to rising concentrations of GHGs. In the coming decades to centuries, the ocean's response to increased GHG concentrations and higher temperatures will continue to alter the climate. Temperatures in the future are anticipated to change considerably. By 2100, average global temperature increases are projected to range from 0.5 to 8.6°F, with a predicted increase of at least 2.7°F for all scenarios save the most aggressive mitigation of greenhouse gas emissions.[7] Land temperatures are anticipated to continue to rise more rapidly than ocean temperatures. Unless the most extreme mitigation option assessed is implemented, the global average temperature is projected to increase by at least twice as much over the next century as it has over the past century.

4.3 CONSEQUENCES OF GLOBAL CLIMATE CHANGE

The dramatic increase in human population since the preindustrial revolution (1850) increased the concentration of these gases (GHGs) because of the large-scale use of fossil fuels, deforestation, industrialization, and urbanization. Anthropogenic activities such as fossil fuel burning and deforestation of land will lead to an exponential rise in the concentration of atmospheric carbon dioxide (CO_2) at the rate of 1 $\mu mol\ mol^{-1}$ by 2050. According to Long et al.,[9] at present, CO_2 is higher than it was anytime in the past 36 million years. Today's crops and natural vegetation are growing at an elevated level of CO_2 that does not seem to have been experienced by terrestrial vegetation in the last 26 million years.[10] Global warming has harmful consequences on the climate and biota, as shown below in the pictorial model (Fig. 4.2).

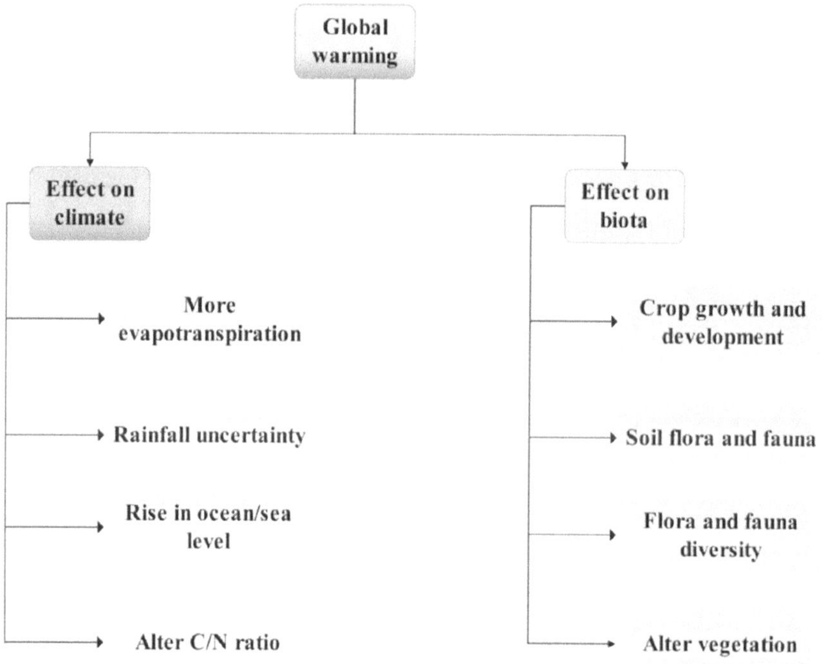

FIGURE 4.2 Effect of global warming on climate and biota.

Climatic alterations have brought about expanded and irregular flood frequencies in the most recent couple of years. An ascent in ocean levels with the predominance of precipitation could bring about the decline of agricultural terrains. This has also resulted in the salinization of soils in beachfront areas, causing yield stresses like decreased crop respiration and photosynthesis and eventually jeopardizing food accessibility and security in such areas. Rising sea levels are also increasing the frequency and severity of flooding on ranches in coastal areas. These expensive floods decimate agricultural harvests, along with soil disintegration, water contamination, etc. Crop yield is dependent on the ideal temperature during the reproductive phase and harvest time.[11] Fluctuations around the ideal temperature cause yield penalties. A dry season, besides temperatures, would dial back or even harm crop development, which would decline crop creation.[12] Increased plant cellular CO_2 accumulation due to elevated atmospheric CO_2 negatively affects protein turnover and essential mineral accumulation.[11]

Environmental change has the potential to disrupt an unsettling influence system that influences the construction and capability of forest ecosystems. Climatic hazards such as fires, disease incidence, and abiotic stress pose a serious threat to forest ecosystem stability. Temperature rises create conditions that significantly elevate the likelihood and magnitude of forest fires. The effects of fire on natural vegetation include the disruption of habitat, the acceleration of biogeochemical cycles, and the mortality of native trees. The extent and frequency of rainfalls, as well as forest structure and composition, influence the likelihood, magnitude, frequency, seasonal patterns, and category of fire. Fire has a significant impact on the backwoods' forest carbon balance and weakens forest carbon sinks.[14] Forest wildfires deplete most soil nutrients, affecting available macronutrient concentrations. Climate change, such as extreme heat and plant biomass productivity, increases the chance of spontaneous fire combustion and propagation, which has an immediate impact on the environment.[15] Rising temperatures are most likely to impact the bioavailability of nutrients in the soil by stimulating organic matter decay and the mineralization of soil nutrients. There are significant connections between trees and soil availability of nutrients. One of the most important aspects of forest ecosystem processes is the circulation of nutrient content between soil and plants. Climate change reduces the availability of nutrient levels for plant uptake, limiting natural forest productivity and survival. The availability of nutrients influences the ecological system and gross ecosystem productivity.[16] Biomass from the forest litter is broken down by microbes in the soil. Increasing nitrogen (N) demand is needed for plant development and protein production; however, as CO_2 levels in the atmosphere rise, more microbes become N-limited, releasing fewer nutrients to trees. The quantity and availability of nutrients appear to control forest ecosystem productivity. Warming temperatures have direct physiological effects on the rate of photosynthesis and the concentration of macronutrients in soils and the leaves of trees, shrubs, and herbs, and temperature increases can cause nutrient availability to decrease.[17]

The level of water stored within the soil is vital to agricultural production because it influences the proportion of existing evaporation, groundwater recharge, and runoff generation. Global climate modeling explicitly approximates soil moisture content, albeit at a very fine spatial resolution, and the outputs of such models indicate future alternatives of transformation. However, the local impacts of global warming based on soil moisture will fluctuate not only based on the degree of climatic changes but also on soil conditions. Soil water retention capacity influences potential shifts in soil hydration imbalances, the lesser the capacity, the more sensitive it is to climate change. Anthropogenic global warming could also have an impact on soil characteristics, such as changes in seasonal flooding or cracking, which may hinder groundwater storage characteristics. The recurrence and severity of the freezing impact the soil permeation abilities and moisture-holding capacity of soils. Erratic precipitation causes alterations in both groundwater recharge and the span of recharge season. A higher rate of evaporation due to the global warming effect puts the soil under water-deficit conditions. The increase in GHGs alters fundamental variables like total carbonate alkalinity (inverse relationship with CO_2 concentration), which creates acidity and carbonate variables in the ocean.[18] Sea level rise is a result of global warming, which accounts for glacier melting and subsequent disturbances in marine and terrestrial biodiversity. Freshwater fauna is threatened by a variety of factors, including nutrient enrichment, hydrogeological changes, biodiversity deterioration, and the proliferation of invasive species. A changing climate and rising levels of light irradiation present emerging threats that amplify existing ones. Rapid urban land use shifts, habitat fragmentation, and global warming are viewed as particularly serious threats to aquatic ecosystems. Freshwater ecosystems are inherently susceptible to hydrogeological cycle changes, which are hard to predict. A warmer climate causes more water loss from surface waters and more stomatal conductance in plants, resulting in a more robust water cycle. Warmer temperatures and changes in hydrological processes will have a direct impact on lake ecosystems in the future.

The influence of GHGs, elevated temperatures, shifting rainfall, and adverse climate incidents on soil microorganisms is direct, while secondary effects result from weather-driven changes in plant productivity and ecosystem processes, which alter soil physicochemical and microbiological conditions, the supply of carbon to the soil, as well as the composition and action of microbiomes associated with the biodegradation process and carbon release from the soil. The temperature rise hastens microbial biodegradation, raising CO_2 emissions by soil respiration and contributing to global warming.[19] Changes in soil temperature are likely to shift the microbiota denitrification and nitrification interplay in the soil environment caused by population shifts of denitrifiers and nitrifiers, respectively.

Food security is one of the concerning scenarios that may lead to compromised food quality, higher food prices, and inadequate food distribution systems. There are more natural disasters and less food security around the world under changing climatic conditions. Several big weather and natural disasters have hurt crop production in the countries in question. Adaptation and mitigation demand special emphasis at both the national and international levels. In recent decades, the world has confronted a significant challenge with climate change, and adaptation to its effects is essential for economic and social progress. To adapt and mitigate climate change, one should develop policies and strategies at the international level.[8] All policy approaches attempt to minimize production costs while increasing the share of renewable energy in a nation's energy system. Long-term contracts with renewable energy providers, government commitment and control, and the setting of long-term goals can help developing nations implement renewable energy technologies in their energy sector.

4.4 NIGHT WARMING EFFECT ON PLANT LIFE STAGES

4.4.1 SEED GERMINATION AND VEGETATIVE STAGE

Physiologically, diurnally alternating temperatures may act through a decrease in the ratio of abscisic acid to gibberellins, which reduces the embryo's water potential and promotes radicle growth.[20–22] The effect of thermal oscillations is almost immediate; seeds undergo a large transcriptome change involving metabolism, plant hormones, and the circadian clock.[23] Increased night temperature may increase the plant leaf's night respiratory rate,[24–26] cell metabolism, and leaf growth rate.[27,28] High respiration during night-time leads to a decrease in the percent of photoassimilates available for plant growth, yield and night respiration rate are linearly correlated with leaf starch and soluble sugar content.[29] Plants respire ~50% of the carbon produced through the photosynthesis process (after photorespiration).[30] However, some results showed that the number of tillers, leaf area, above-ground weight, and root dry weight per plant increased because of high night temperature (HNT), but the crop growth duration decreased, as indicated by earlier panicle emergence.[31] The reason behind increasing the number and growth of vegetative structures is a decrease in pollen germination, which reduces the number of reproductive sinks, leading to an increase in dry matter partitioning into vegetative structures.[32] Carbon accumulation and plant growth are affected by plant photosynthesis and night respiration. Night warming at the vegetative stage promotes the postanthesis accumulation of dry matter and the transfer of the dry matter to grains, which ultimately benefits the yield. Stomata play a key role in the exchange of gases and water between plants and their surroundings. The quantity, size, and regulatory role of stomata were associated with physiological activities including photosynthesis and transpiration of leaves, which have a significant impact on crop productivity and tolerance to water stress and temperature stress.[33] Under night warming conditions, the stomatal density and stomatal index of flag leaves significantly increase. The spatial arrangement of plant leaf stomata is determined by the processes of cell division and differentiation. Regulation of environmental conditions and genetic signals might have an impact on cell division and differentiation.[34]

4.4.2 REPRODUCTION AND FLOWERING STAGE

Early reproductive development is most likely the parameter of yield, which is more sensitive than vegetative development to transient high-temperature events.[35] High air temperatures lose pollen viability more rapidly due to the exponential increase in vapor pressure deficit with temperature, and a 50% loss in viability might occur. HNT increases pollen reactive oxygen species (ROS), whose elevated concentrations cause oxidative damage and cell death,[36,37] which leads to membrane damage and decreases pollen functions such as viability and germination.[38] In the case of wheat spikelets and florets, differentiation occurs during the tillering to booting stages; there is a high probability of winter wheat suffering freeze injury in these stages. Night warming treatment during this period, therefore, benefits the development of spikelets and florets.[39]

4.4.3 YIELD AND QUALITY

The impact of high temperatures on plant metabolism depends on how long and how intensely plants are exposed to them, which are between 10 and 15°C above ambient.[40] However, depending on the crop's phenological stage at the time of exposure, the effects on economic yield vary.[41] Dry matter is the product of photosynthesis, and dry matter accumulation, distribution, and translocation decide the grain yield.[42] Increased respiration during night warming enhances carbohydrate consumption, stimulating daytime photosynthesis and thus increasing dry matter accumulation.[43] The formation and accumulation of dry matter during the early stages of grain development are facilitated by night warming because it increases the net photosynthetic rate of wheat flag leaves at the early stage of grain filling, increases the chlorophyll content of flag leaves after anthesis, and effectively improves the maximum photochemical efficiency and actual photochemical efficiency of wheat flag leaves after anthesis.[39] Warming before anthesis alleviates the negative effects on wheat from high temperatures by enhancing flag leaf photosynthesis and dry matter remobilization from stems to grains after anthesis.[44,45] Plants experiencing night warming at early growth stages have a higher amount of dry matter accumulation after anthesis and a higher harvest index.[39] However, in some crops, HNT increases carbon loss from night respiration and reduces the activity of source–sink enzymes, resulting in lower biomass, poor grain filling, and reduced grain weight.[46] In field-grown rice, night temperature-driven reductions in grain yield extend to poor grain quality and increase chalkiness,[47] as a result of a reduction in major starch synthesizing enzymes. In the case of wheat, however, the decrease in starch concentration driven by carbon losses due to increasing nighttime respiration led to a greater accumulation of protein and lipids,[48] but such gains, particularly in protein, were not enough to compensate for the incurred yield penalties. The effect of elevated night temperatures is highlighted in Figure 4.3.

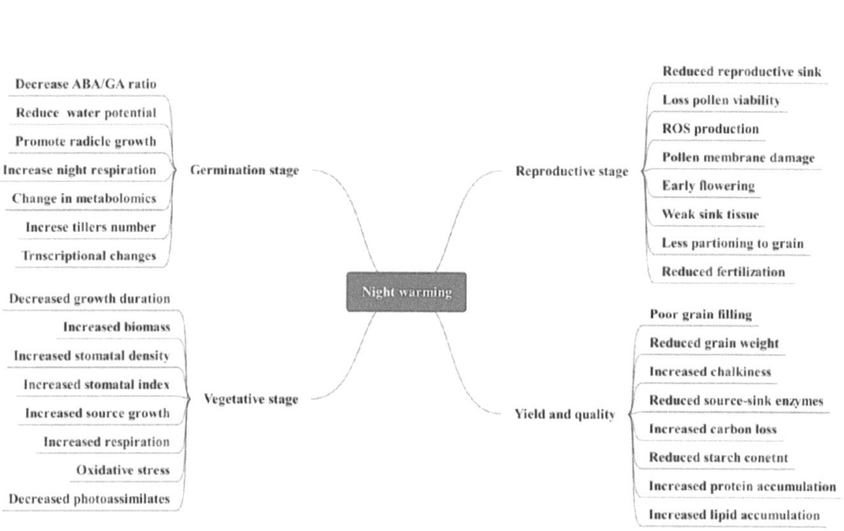

FIGURE 4.3 Effect of night warming on the different stages of plant growth.

4.5 NIGHT WARMING EFFECT ON THE CRUCIAL PROCESS

4.5.1 PHOTOSYNTHESIS

The continuous rise of mean temperature is one of the most likely phenomena occurring in the 21st century. Several climate analyses have explained that the unique feature of this global warming is an uneven increase in daily extreme temperatures,[49] as the night temperature (i.e., minimum temperature) has increased at a faster rate than the day temperature (i.e., maximum temperature).[50] HNTs increase crop respiration, which is responsible for up to half of the carbon loss from fixed carbon per day photosynthesis.[51,52] Several shreds of evidence from previous reports have presented that HNT decreased photosynthesis efficiency in wheat[53] and increased respiration.[54] Moreover, few reports suggest that HNT can affect plants by influencing the vapor pressure deficit (VPD). This component is very new, even though the agrometrological data are not yet well documented for night-time conditions, and it is considered that elevated night temperatures will be linked with increases in night-time VPD, regardless of the environments of cultivated crops.[55] Stomata closure is linked with high VPD, which reduces photosynthetic efficiency.[56] The photosynthetic efficiency can be reduced by elevated night temperatures through the influence on photosystems,[57] stomatal conductance, internal CO_2 concentration,[58] enzymes involved in carbon fixation,[59] and leaf N concentration.[31] Another group of mechanisms that imitate the effects of higher night temperatures on photosynthesis is decreased maximum efficiency of photosystem II (Fv/Fm) and membrane damage due to the accumulation of ROS. This ROS remarkably disrupts net photosynthesis by disrupting principally target enzymes involved in carbon fixation if it is not scavenged at a time. Thus, previous reports suggested that the biochemical mechanism of photosynthesis is severely ruptured by ROS in leaves in HNTs. In line with this, new reports predict that photosynthetic efficiency reduction may also arise due to the deactivation of several key chloroplastic stroma enzymes during the night,[48] as a consequence of quicker rates of leaf senescence.[60] Furthermore, very few evidence points to a role of night warming in downregulating photosynthesis the next day, probably through a pathway of feedback inhibition due to the overaccumulation of carbohydrates in the leaves.[61]

4.5.2 NIGHT WARMING EFFECT ON TRANSLOCATION IN PLANTS

HNT in vegetative tissues leads to yield reduction by affecting the carbon amount, which is directed toward the growth of seeds and plants. Night warming induces an enhancement in respiration at night, which results from the decrease in photoassimilates' amount available for growth and yield of plants.[24, 46, 62] It has been studied that a reduced rate of translocation of N and nonstructural carbohydrates (NSC) decreases the rate of grain filling, which affects the weight and quality of rice grain.[46] Under rising temperatures, there is evidence of enhanced respiratory activity in plants that is thermodynamically driven, which may be responsible for a 30–80% reduction in carbon assimilation.[24,62,63] On the other hand, recent studies suggested that night warming could change the rate of translocation and accumulation of photoassimilates during the grain-filling stage.[46,64] Night respiration did not change significantly during the preflowering phase. This might be attributed to night respiration showing low sensitivity to HNT during the vegetative crop growth phase, which has been linked to a lower percentage of respiratory requirements for sustaining growing biomass.[65,66] Night respiration, on the other hand, increased dramatically after the flowering stage, which is associated with enhanced translocation of photoassimilates to developing seeds, the enzymatic activity of the sink for the unloading of sugar, and the accumulation of starch for seed storage.[67–69] The primary transportable sugar in plants is sucrose, which is moved from source to sink through the gradient of sucrose and energy-driven phloem transport.[70,71] According to the study of Bahuguna et al.,[25] total soluble sugars accumulated more in the rachis tissue that was connected to the panicle during the stage of grain filling, and the sugar content of the spikelets decreased under HNT. By cleaving sucrose in the apoplast, cell wall invertase (CWI), which is an insoluble protein ionically linked to cell walls, plays a crucial role in phloem unloading.[72] Lower CWI activity noted for Gharib and IR64 provided preliminary evidence for HNT impact on phloem unloading.[25] A comparative study of two rice genotypes, N22 (high temperature tolerant) and Gharib (high temperature sensitive), at 22 and 28°C temperatures, independently revealed a smaller decrease in NSC translocation in Gharib, which could be attributed to the HNT effect alone, whereas a larger decrease in N22 could be caused by a combination of the HNT effect and reduced sink size.[46]

4.5.3 NIGHT WARMING EFFECT ON MINERAL NUTRITION IN PLANTS

High night-time temperatures may promote soil mineralization of N and phosphorus (P), which could alter the nutrient intake of plant roots. One of the key factors affecting plant production and photosynthetic capacity is nutrient (N and P) uptake. Numerous research studies have demonstrated that warming may either increase or reduce the N accumulation of the examined plants, the majority of which were carried out in natural ecosystems without N fertilizer.[73–75] An important factor in plant N uptake is root development, which can be enhanced to boost plant N uptake and decrease nitrate leakage into the environment. Some researchers reported the effect of night warming on the morphology and activity of plant roots and showed in coniferous species that night warming increases root length, root biomass, and the root/shoot mass ratio,[76–79] and also that winter wheat's root activity and dry root biomass increase with warming, which enhances plant N intake.[28] At high temperatures (+4°C), leaf photorespiration and dark respiration are inhibited, which reduces nitrate reductase (NR) and glutamine synthetase (GS) activity in wheat leaves. The night warming treatments in this study, however, increased NR and GS activity in leaves at the jointing, booting, anthesis, and filling stages. Wheat N accumulation was enhanced as a result of the leaves higher NR and GS activity, which also improved N assimilation capacity.[80] Increases in VPD during the night represent an even less well-known aspect of night-time warming. There is evidence of significant and detrimental consequences of a rise in night VPD on calcium uptake and export to plant tissues, including reproductive organs. This is likely because calcium naturally translocates at greater rates at night than it does during the day.[81] According to studies, a relatively small increment in night VPD, from 0.26 to 0.76 kPa, resulted in a calcium (Ca) content drop in tomatoes of up to 66%.[82] Similar results were observed in lettuce (*Lactuca sativa*), where increasing night VPD from 0.07 to 0.15 kPa caused a drop in leaf Ca of almost 33%.[83] The uptake of P and zinc (Zn) by *M. truncatula* was negatively impacted by the slight, night-time steady temperature increase (1.53°C). As a result, plant nutrient uptake may be extremely sensitive to changes in surrounding temperature, such as warming and cooling. Zn, which serves as a crucial enzyme cofactor in the control of microsporogenesis and pollen fertility, is a crucial micronutrient for plant reproduction.[84] Studies show that Zn contents in plant shoots dramatically decreased during night warming.[85]

4.5.4 METABOLISM

The asymmetric rising temperature over the decades is very common in the present global warming era due to increased cloudiness, reduced back radiation, air pollution by vehicles and aerosols, and changes in precipitation patterns.[86] It could intensify the manifestation of new abiotic stress on plants, such as HNT stress, which could alter the plant metabolism, resulting in potential effects on a wide range of growth and developmental stages.[87] Previous studies have shown that HNT increased respiration[54] in different crops; thereby, carbon and N metabolism flow is diverted to the Krebs cycle and glycolysis. During respiration, drastic losses in the carbon amount of different metabolic processes under HNT, which have shown a negative impact on photosynthetic machinery and hampered the resulting photosynthetic rate, were documented in rice under field conditions.[25] During night warming, initially in the dark period, CO_2 efflux takes place; hence, CO_2 balance is hampered.[88] It also influenced PEPC activity and CO_2 fixation, resulting in the C_4 and CAM plants' carbon metabolism routes diverting compared to the normal cycle. Under high night stress, sink (spike) metabolic changes are more as compared to leaf and stem metabolites due to less supply of soluble carbohydrates to sink during grain filling.[89] In general, in sink-limited crops, HNT accelerates the starch degradation of enzymes, that is, isoamylases and amylases.[48] Starch degradation occurs to maintain the high respiratory standards. It increased the accumulation of soluble sugar (raffinose and maltose), glucose, fructose, and sugar alcohol, that is, myoinositol,[90] in leaves, which has to lack the capacity to recycle transportable sugar phosphate end products due to improper translocation.[61] A recent study reports that the soluble sugar acts as an osmoprotectant and is also involved in neutralizing the ROS to withstand night warming.[48,90,91] The metabolic analysis depicts that carbon losses in the form of nonstructural or structural carbohydrates in grape vines and cereals reflect the high consumption, boosting respiration.[61] It deciphers that HNT, due to respiration, enhanced the metabolic flow in the direction of the Krebs cycle at α keto-glutaric acid, isocitrate, and fumarate oxaloacetic acid to support increased respiration in the photosynthesizing tissue.[48,90,92] The TCA cycle intermediate compound goes through another pathway, that is, the shikimate pathway, through which accumulation of several stress-induced protein biosynthesis and breakdown occurs. In the case of N metabolism, uptake is greatly enhanced under HNT. The process of ammonia assimilation by key enzymes GS and glutamate dehydrogenase (GDH) is also altered by the N level in rice. The synthesis of aromatic (tyrosine) and aspartate-derived amino acids such as leucine, lysine, methionine, and proline increased, as well as some protein involved in protein degradation were also increased. For example, degradation of Rubisco associate with leaf senescence,[93] and the synthesis of new proteins enhances physiological tolerance to HNT.[94–98] The effect of night warming on crucial processes is highlighted in Figure 4.4.

4.6 CONCLUSIONS

One of the major driving forces for global climate change is the exponential increase in concentrations of GHGs in the atmosphere. In the coming decades to centuries, the ocean's response to increased GHG concentrations and higher temperatures will continue to alter the global climate. Natural hazards resulting from global climate change cause the incidence of stresses like drought, submergence, and temperature concerns, for which the plants

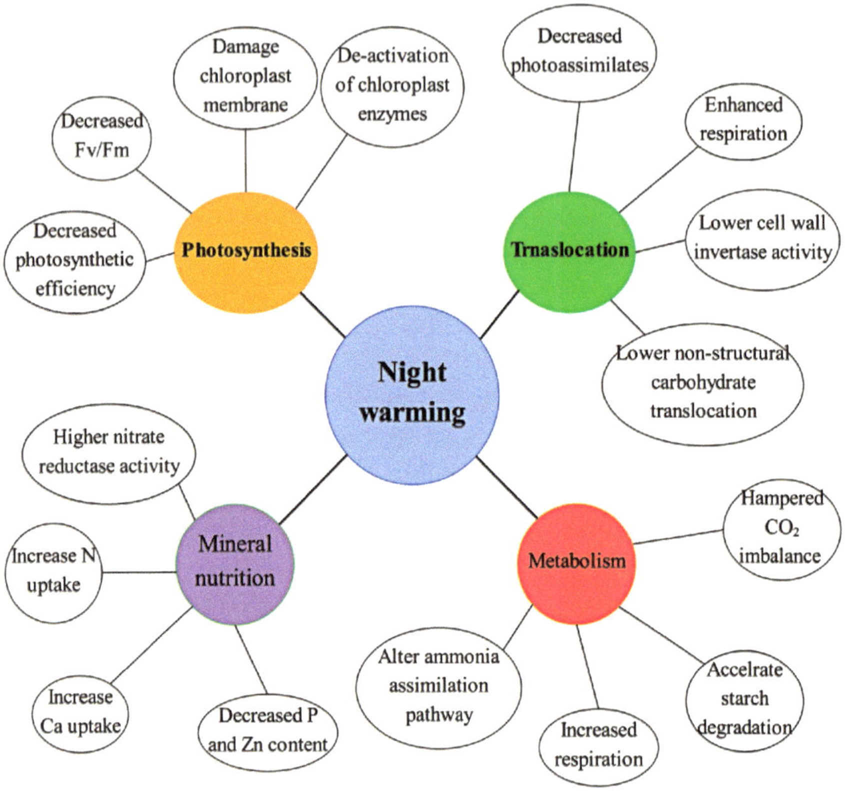

FIGURE 4.4 Effect of elevated night temperature on crucial plant processes.

must undergo severe physiological and metabolic alterations, including shuffling in primary and secondary metabolite production and their organelle distribution. These adjustments affect the plant's capacity to meet the expected yield attributes. Crucial plant growth and developmental phases such as seed germination, seedling growth, vegetative, reproductive, and maturity phases might be severely affected by night-time warming. These growth phases are regulated by photoassimilates' capacity, metabolism, partitioning, and transport to sink tissues, which ultimately affect the yield and quality of crop products. Therefore, it is crucial to understand the in-depth physiological, biochemical, and molecular mechanisms or responses under the control of night-time warming.

KEYWORDS

- **climate change**
- **photosynthesis**
- **metabolism**
- **yield penalty**
- **respiration**

REFERENCES

1. Leal Filho, W.; Azeiteiro, U. M.; Balogun, A. L.; Setti, A. F. F.; Mucova, S. A.; Ayal, D.; Oguge, N. O. The inflence of Ecosystems Services Depletion to Climate Change Adaptation efforts in Africa. *Sci. Total Environ.* **2021**, 146414.
2. Feliciano, D.; Recha, J.; Ambaw, G.; MacSween, K.; Solomon, D.; Wollenberg, E. Assessment of Agricultural Emissions, Climate Change Mitigation and Adaptation Practices in Ethiopia. *Clim. Policy* **2022**, 1–18.
3. Weisheimer, A.; Palmer, T. Changing Frequency of Occurrence of Extreme Seasonal Temperatures Under Global Warming. *Geophys. Res. Lett.* **2005**, *32* (20).
4. Yadav, M. K.; Singh, R.; Singh, K.; Mall, R.; Patel, C.; Yadav, S.; Singh, M. Assessment of Climate Change Impact on Productivity of Different Cereal Crops in Varanasi. India *J. Agrometeorol.* **2015**, *17* (2), 179–184.
5. Schuurmans, C. The World Heat Budget: Expected Changes Climate Change; CRC Press, 2021; pp 1–15.
6. Nicholls, N.; Gruza, G. V.; Jouzel, J.; Karl, T. R.; Ogallo, L. A.; Parker, D. E. Observed Climate Variability and Change–; Cambridge University Press: Cambridge, 1996; pp 133–192.
7. Stocker, T. F.; D. Qin, G. K.; Plattner, M.; Tignor, S. K.; Allen, J.; Boschung, A.; Nauels, Y.; Xia, V.; Midgley P. M. Climate Change 2013: The Physical Science Basis Contribution of Working Group I to the Fifth Assessment Report of the Intergovernmental Panel on Climate, **2013**.
8. Hussain, M.; Butt, A. R.; Uzma, F.; Ahmed, R.; Irshad, S.; Rehman, A.; Yousaf, B. A Comprehensive Review of Climate Change Impacts, Adaptation, and Mitigation on Environmental and Natural Calamities in Pakistan. *Environ. Monit. Assess.* **2020**, *192* (1), 48.
9. Long, S. P.; Ainsworth, E. A.; Rogers, A.; Ort, D. R. Rising Atmospheric Carbon Dioxide: Plants FACE the Future. *Annu. Rev. Plant Biol.* **2004**, *55*, 591–628.
10. Pearson, P. N.; Palmer, M. R. Atmospheric Carbon Dioxide Concentrations Over the Past 60 Million Years. *Nature* **2000**, *406* (6797), 695–699.
11. Hatfield, J.; Takle, G.;Grotjahn, R.; Holden, P.; Izaurralde, R. C.; Mader, T.; Marshall, E.; Liverman, D. Ch. 6: Agriculture: Climate Change Impacts in the United States:

The Third National Climate Assessment; Melillo, J. M., Richmond, Terese (T. C.), Yohe, G. W., Eds.; U.S. Global Change Research Program, 2014; pp 150–174.
12. Mendelsohn, R. The Impact of Climate Change on Agriculture in Asia. *J. Integ. Agric.* **2014**, *13* (4), 660–665.
13. Dawadi, B.; Shrestha, A.; Acharya, R. H.; Dhital, Y. P.; Devkota, R. Impact of Climate Change on Agricultural Production: A Case of Rasuwa District, Nepal. *Regional Sustainability* **2022**, *3* (2), 122–132.
14. Gebeyehu, M. N.; Hirpo, F. H. Review on Effect of Climate Change on Forest ecosystem. *Int. J. Environ. Sci. Nat. Resour.* **2019**, *17* (4), 126–129.
15. Tyukavina, A.; Potapov, P.; Hansen, M. C.; Pickens, A. H.; Stehman, S. V.; Turubanova, S.; Harris, N. Global Trends of Forest Loss Due to Fire from 2001 to 2019. *Front. Remote Sens.* **2022**, *3*.
16. Dukes, J. S.; Classen, A. T.; Wan, S.; Langley, J. A. Using Results from Global Change Experiments to Inform Land Model Development and Calibration. *New Phytol.* **2014**, *204* (4), 744–746.
17. Juneja, A.; Ceballos, R. M.; Murthy, G. S. Effects of Environmental Factors and Nutrient Availability on the Biochemical Composition of Algae for Biofuels Production: A Review. *Energies* **2013** *6* (9), 4607–4638.
18. Bach, L. T.; Gill, S. J.; Rickaby, R. E.; Gore, S.; Renforth, P. CO_2 Weathering Removal and with Ocean Enhanced Alkalinity Enhancement: Potential Risks and Co-Benefits for marine Pelagic ecosystems. The Role of Negative Emission Technologies in Addressing Our Climate Goals, 2020.
19. Bölscher, T.; Ågren, G. I.; Herrmann, A. M. Land-Use Alters the Temperature Response of Microbial Carbon-Use Efficiency in Soils–A Consumption-Based Approach. *Soil Biol. Biochem.* **2020**, *140*, 107639.
20. Huarte, H. R.; Benech-Arnold, R. L. Incubation Under Fluctuating Temperatures Reduces Mean Base Water Potential for Seed Germination in Several Non-Cultivated Species. *Seed Sci. Res.* **2005**, 15, 89–97.
21. Huarte, H. R.; Benech-Arnold, R. L. Hormonal Nature of Seed Responses to Fluctuating Temperatures in *Cynara cardunculus* (L.). *Seed Sci. Res.* **2010**, *20*, 39–45.
22. Huarte, H. R. *Mecanismos fisiolo´gicos y moleculares involucrados en la terminacio´n de la dormicio´n de semillas expuestas a las temperaturas alternadas*; PhD thesis, University of Buenos Aires, 2013.
23. Foley, M. E.; Anderson, J. V.; Chao, W. S.; Dog˘ramacı, M.; Horvath, D. P. Initial Changes in the Transcriptome of Euphorbia Esula Seeds Induced to Germinate with a Combination of Constant and Diurnal Alternating Temperatures. *Plant Mol. Biol.* **2010**, *73*, 131–142.
24. Atkin, O. K.; Bruhn, D.; Hurry, V. M.; Tjoelker, M. G. Evans Review No. 2: The Hot and the Cold: Unravelling the Variable Response of Plant Respiration to Temperature. *Funct. Plant Biol.* **2005**, *32*, 87–105.
25. Bahuguna, R. N.; Solis, C. A.; Shi, W.; Jagadish, K. S. Post-Flowering Night Respiration and Altered Sink Activity Account for High Night Temperature-Induced Grain Yield and Quality Loss in Rice (*Oryza sativa* L.). *Physiol. Plantar.* **2017**, *159* (1), 59–73.

26. García, G. A.; Dreccer, M. F.; Miralles, D. J.; Serrago, R. A. High Night Temperatures During Grain Number Determination Reduce Wheat and Barley Grain Yield: A Field Study. *Global Chang. Biol.* **2015**, *21* (11), 4153–4164.
27. Chen, J.; Tian, Y.; Zhang, X.; Zheng, C.; Song, Z.; Deng, A.; Zhang, W. Night-time Warming Will Increase Winter Wheat Yield Through Improving Plant Development and Grain Growth in North China. *J. Plant Growth Regul.* **2014**, *33*, 397–407.
28. Tian, Y.; Zheng, C.; Chen, J.; Chen, C.; Deng, A.; Song, Z.; Zhang, W. Climatic Warming Increases Winter Wheat Yield But Reduces Grain Nitrogen Concentration in East China. *PloS One* **2014**, *9* (4), e95108.
29. Peraudeau, S.; Lafarge, T.; Roques, S.; Quiñones, C. O.; Clement-Vidal, A.; Ouwerkerk, P. B.; Dingkuhn, M. Effect of Carbohydrates and Night Temperature on Night Respiration in Rice. *J. Exp. Bot.* **2015**, *66* (13), 3931–3944.
30. Ryan, M. G.; Lavigne, M. B.; Gower, S. T. Annual Carbon Cost of Autotrophic Respiration in Boreal Forest Ecosystems in Relation to Species and Climate. *J. Geophy. Res. Atm.* **1997**, *102*, 28871–28883.
31. Mohammed, A. R.; Tarpley, L. High Night-time Temperature Affects Rice Productivity Through Altered Pollen Germination and Spikelet Fertility. *Agric. For. Meteorol.* **2009**, *149*, 999–1008. DOI: 10.1016/j.agrformet.2008.12.003.
32. Prasad, P. V. V.; Boote, K. J.; Allen Jr., L. H.; Sheehy, J. E.; Thomas, J. M. G. Species, Ecotype and Cultivar Differences in Spikelet Fertility and Harvest Index of Rice in Response to High Temperature Stress. *Field Crops Res.* **2006**, *95*, 398–411.
33. Casson, S.; Gray, J. E. Influence of Environmental Factors on Stomatal Development. *New Phytol.* **2008**, *178*, 9–23.
34. Haworth, M.; Marino, G.; Loreto, F.; Centritto, M. Integrating Stomatal Physiology and Morphology: Evolution of Stomatal Control and Development of Future Crops. *Oecologia* **2021**, *197*, 867–883.
35. Siebers, M. H.; Slattery, R. A.; Yendrek, C. R.; Locke, A. M.; Drag, D.; Ainsworth, E. A.; Ort, D. R. Simulated Heat Waves During Maize Reproductive Stages Alter Reproductive Growth But Have No Lasting Effect When Applied During Vegetative Stages. *Agric. Ecosys. Environ.* **2017**, *240*, 162–170.
36. Apel, K.; Hirt, H. Reactive Oxygen Species: Metabolism, Oxidative Stress, and Signal Transduction. *Ann. Rev. Plant Biol.* **2004**, *55*, 373–399.
37. Djanaguiraman, M.; Prasad, P. V. V.; Seppanen, M. Selenium Protects Sorghum Leaves from Oxidative Damage Under High Temperature Stress by Enhancing Antioxidant Defense System. *Plant Physiol. Biochem.* **2010**, *48*, 999–1007.
38. Prasad, P. V.; Djanaguiraman, M. High Night Temperature Decreases Leaf Photosynthesis and Pollen Function in Grain Sorghum. *Funct. Plant Biol.* **2011**, *38* (12), 993–1003.
39. Fan, Y.; Lv, Z.; Zhang, Y.; Ma, L.; Qin, B.; Liu, Q.; Huang, Z.; Pre-Anthesis Night Warming Improves Post-Anthesis Physiological Activity and Plant Productivity to Post-Anthesis Heat Stress in Winter Wheat (*Triticum aestivum* L.). *Environ. Exp. Bot.* **2022**, *197*, 104819.
40. Wahid, A.; Gelani, S.; Ashraf, M.; Foolad, M. R. Heat Tolerance in Plants: An Overview. *Environ. Exp. Bot.* **2007**, *61*, 199–223.
41. Hall, A. E. Breeding for Heat Tolerance. *Plant Breed. Rev.* **1992**, *10* (2), 129–168.

42. Masoni, A.; Ercoli, L.; Mariotti, M.; Arduini, I. Post-Anthesis Accumulation and Remobilization of Dry Matter, Nitrogen and Phosphorus in Durum Wheat as Affected by Soil Type. *Euro. J. Agron.* **2007**, *26* (3), 179–186.
43. Way, D. A.; Yamori, W. Thermal Acclimation of Photosynthesis: On the Importance of Adjusting Our Definitions and Accounting for Thermal Acclimation of Respiration. *Photosyn. Res.* **2014**, *119*, 89–100.
44. Wang, X.; Cai, J.; Jiang, D.; Liu, F.; Dai, T.; Cao, W.; Pre-Anthesis High-Temperature Acclimation Alleviates Damage to the Flag Leaf Caused by Post-Anthesis Heat Stress in Wheat. *J. Plant Physiol.* **2011**, *168*, 585–593.
45. Wang, X.; Cai, J.; Liu, F.; Jin, M.; Yu, H.; Jiang, D.; Wollenweber, B.; Dai, T.; Cao, W. Pre-Anthesis High Temperature Acclimation Alleviates the Negative Effects of Post-Anthesis Heat Stress on Stem Stored Carbohydrates Remobilization and Grain Starch Accumulation in Wheat. *J. Cereal Sci.* **2012**, *55*, 331–336.
46. Shi, W.; Muthurajan, R.; Rahman, H.; Selvam, J.; Peng, S.; Zou, Y.; Jagadish, K. S. Source–Sink Dynamics and Proteomic Reprogramming Under Elevated Night Temperature and Their Impact on Rice Yield and Grain Quality. *New Phytol.* **2013**, *197* (3), 825–837.
47. Lyman, N. B.; Jagadish, K. S.; Nalley, L. L.; Dixon, B. L.; Siebenmorgen, T.; Neglecting Rice Milling Yield and Quality Underestimates Economic Losses from High-Temperature Stress. *PloS One* **2013**, *8* (8), e72157.
48. Impa, S. M.; Sunoj, V. J.; Krassovskaya, I.; Bheemanahalli, R.; Obata, T.; Jagadish, S. K. Carbon Balance and Source-Sink Metabolic Changes in Winter Wheat Exposed to High Night-Time Temperature. *Plant Cell Environ.* **2019**, *42* (4), 1233–1246.
49. Dunn, R. J.; Alexander, L. V.; Donat, M. G.; Zhang, X.; Bador, M.; Herold, N.; Bin, H. J.; Yussof, M. N. A. Development of an Updated Global Land In Situ-Based Data Set of Temperature and Precipitation Extremes: HadEX3. *J. Geophys. Res.: Atmos.* **2020**, *125* (16), e2019JD032263.
50. Vose, R. S.; Easterling, D. R.; Gleason, B. Maximum and Minimum Temperature Trends for the Globe: An Update Through 2004. *Geophy. Res. Lett.* **2005**, *32*, L23822. https://doi.org/10.1029/2005GL024379.
51. Atkin, O. K.; Scheurwater, I.; Pons, T. L. Respiration as a Percentage of Daily Photosynthesis in Whole Plants Is Homeostatic at Moderate, But Not High, Growth Temperatures. *New Phytol.* **2007**, *174*, 367–380. https://doi.org/10.1111/j.1469-8137.2007.02011.x.
52. Alwyn, W.; Kylie, W.; Millicent Smith, R. C. N.; Rachaputi; Marisa, C. Mungbean Responses to Elevated Day-and Night-Time Temperatures. *Proceedings of the 20th Agronomy Australia Conference*, 2022, Toowoomba. www.agronomyaustraliaproceedings.org.
53. Prasad, P. V. V.; Pisipati, S. R.; Ristic, Z.; Bukovnik, U.; Fritz, A. K. Impact of Night-time Temperature on Physiology and Growth of Spring Wheat. *Crop Sci.* **2008**, *48* (6), 2372–2380. https://doi. org/10.2135/cropsci2007.12.0717.
54. Mohammed, A. R.; Tarpley. L. Impact of High Night-time Temperature on Respiration, Membrane Stability, Antioxidant Capacity, and Yield of Rice Plants. *Crop Sci.* **2009**, *49*, 313–322. doi:10.2135/cropsci2008.03.0161.

55. Groh, J.; Pütz, T.; Gerke, H. H.; Vanderborght, J.; Vereecken, H. Quantification and Prediction of Night-time Evapotranspiration for Two Distinct Grassland Ecosystems. *Water Resour. Res.* **2019**, *55* (4), 2961–2975.
56. Grossiord, C.; Buckley, T. N.; Cernusak, L. A.; Novick, K. A.; Poulter, B.; Siegwolf, R. T. W.; Sperry, J. S.; McDowell, N. G. Plant Responses to Rising Vapor Pressure Deficit. *New Phytol.* **2020**, *226*, 1550–1566. https://doi.org/10.1111/nph.16485.
57. Guo, Y. P., Zhou, H. F.; Zhang, L. C. Photosynthetic Characteristics and Productive Mechanism Against Photooxidation During High Temperature Stress in Two Citrus Species. *Sci. Hortic.* (Amsterdam) **2006**, *108*, 260–267. DOI: 10.1016/j.scienta.2006.01.029.
58. Farquhar, G. D.; Sharkey T. D. Stomatal Conductance and Photosynthesis. *Ann. Rev. Plant Physiol.* **1982**, *33*, 317–345. DOI: 10.1146/annurev.pp.33.060182.001533.
59. Noctor, G.; Foyer, C. H. Ascorbate and Glutathione: Keeping Active Oxygen Under Control. *Ann. Rev. Plant Physiol. Mol. Biol.* **1998**, *49*, 249–279. DOI: 10.1146/annurev.arplant.49.1.249.
60. Lesjak, J.; Calderini, D. F. Increased Night Temperature Negatively Affects Grain Yield, Biomass and Grain Number in Chilean Quinoa. *Front. Plant Sci.* **2017**, *8*, 352.
61. Tombesi, S.; Cincera, I.; Frioni, T.; Ughini, V.; Gatti, M.; Palliotti, A.; Poni, S. Relationship Among Night Temperature, Carbohydrate Translocation and Inhibition of Grapevine Leaf Photosynthesis. *Environ. Exp. Bot.* **2019**, *157*, 293–298.
62. Atkin, O. K.; Tjoelker, M. G. Thermal Acclimation and the Dynamic Response of Plant Respiration to Temperature. *Trends Plant Sci.* **2003**, *8*, 343–351.
63. Liang, J.; Xia, J.; Liu, L.; Wan, S. Global Patterns of the Responses of Leaf-Level Photosynthesis and Respiration in Terrestrial Plants to Experimental Warming. *J. Plant Ecol.* **2013**, *6*, 437–447.
64. Loka, D. A.; Oosterhuis, D. M. Effect of High Night Temperatures on Cotton Respiration, ATP Levels and Carbohydrate Content. *Environ. Exp. Bot.* **2010**, *68*, 258–263.
65. Frantz, J. M.; Cometti, N. N.; Bugbee, B. Night Temperature Has a Minimal Effect on Respiration and Growth in Rapidly Growing Plants. *Ann. Bot.* **2004**, *94*, 155–166.
66. Rajeev, N.; Celymar, A.; Krishna, S. V. Postâ Flowering Night Respiration and Altered Sink Activity Account for High Night Temperatureâ Induced Grain Yield and Quality Loss in Rice (*Oryza sativa* L.). *Physiol. Plantar.* **2017**, *159*, 59–73.
67. Amthor, J. S. The Role of Maintenance Respiration in Plant Growth. *Plant Cell Environ.* **1984**, *7*, 561–569
68. Amthor, J. S. The McCree–de Wit–Penning de Vries–Thornley Respiration Paradigms: 30 Years Later. *Ann. Bot.* **2000**, *86*, 1–20.
69. Ryan, M. G. Effects of Climate Change on Plant Respiration. *Eco. App.* **1991**, *1*, 157–167.
70. Sturm, A.; Tang, G. Q. The Sucrose-Cleaving Enzymes of Plants Are Crucial for Development, Growth and Carbon Partitioning. *Trends Plant Sci.* **1999**, *4*, 401–407.
71. Koch, K. Sucrose Metabolism: Regulatory Mechanisms and Pivotal Roles in Sugar Sensing and Plant Development. *Curr. Opi. Plant Biol.* **2004**, *7*, 235–246.
72. Wang, L.; Ruan, Y. L. New Insights into Roles of Cell Wall Invertase in Early Seed Development Revealed by Comprehensive Spatial and Temporal Expression Patterns of GhCWIN1 in Cotton. *Plant Physiol.* **2012**, *160*, 777–787.

73. Sardans, J.; Peñuelas, J.; Estiarte, M. Seasonal Patterns of Root-Surface Phosphatase Activities in a Mediterranean Shrubland: Responses to Experimental Warming and Drought. *Biol. Fert. Soil* **2007**, *43*, 779–786.
74. Patil R. H.; Laegdsmand, M.; Olesen, J. E.; Porter J. R. Effect of Soil Warming and Rainfall Patterns on Soil N Cycling in Northern Europe. *Agric. Econ. Environ.* **2010**, *139*, 195–205.
75. Zhang, Y.; Li, R.; Wang, Y. Night-Time Warming Affects N and P Dynamics and Productivity of Winter Wheat Plants. *Canadian J. Plant Sci.* **2013,** *93* (3), 397–406.
76. Björk, R. G.; Majdi, H.; Klemedtsson, L.; Lewis-Jonsson, L.; Molau, U.; Long-Term Warming Effects on Root Morphology, Root Mass Distribution, and Microbial Activity in Two Dry Tundra Plant Communities in Northern Sweden. *New Phytol.* **2007**, *176* (4), 862–873.
77. Bai, W.; Wan, S.; Niu, S.; Liu, W.; Chen, Q.; Wang, Q.; Li, L. Increased Temperature and Precipitation Interact to Affect Root Production, Mortality, and turnover in a Temperate Steppe: Implications for Ecosystem C Cycling. *Glob. Chang. Biol.* **2010**, *16* (4), 1306–1316.
78. Yin, H.; Chen, Z.; Liu, Q. Effects of Experimental Warming on Soil N Transformations of Two Coniferous Species, Eastern Tibetan Plateau, China. *Soil Biol. Biochem.* **2012**, *50*, 77–84.
79. Yin, H.; Xiao, J.; Li, Y.; Chen, Z.; Cheng, X.; Zhao, C.; Liu, Q. Warming Effects on Root Morphological and Physiological Traits: The Potential Consequences on Soil C Dynamics as Altered Root Exudation. *Agric. Fore. Meteor.* **2013**, *180*, 287–296.
80. Hu, C.; Tian, Z.; Gu, S.; Guo, H.; Fan, Y.; Abid, M.; Dai, T. Winter and Spring Night-Warming Improve Root Extension and Soil Nitrogen Supply to Increase Nitrogen Uptake and Utilization of Winter Wheat (*Triticum aestivum* L.). *Eur. J. Agron.* **2018**, *96*, 96–107.
81. Tachibana, S. Import of Calcium by Tomato Fruit in Relation to the Day-Night Periodicity. *Sci. Hortic.* **1991**, *45* (3–4), 235–243.
82. Bradfield, E. G.; Guttridge, C. G. Effects of Night-Time Humidity and Nutrient Solution Concentration on the Calcium Content of Tomato Fruit. *Sci. Hortic.* **1984**, *22* (3), 207–217.
83. Collier, G. F.; Tibbitts, T. W. Effects of Relative Humidity and Root Temperature on Calcium Concentration and Tipburn. *J. Am. Soc. Hort. Sci.* **1984**, *109* (2), 128–131.
84. Pandey, N.; Pathak, G. C.; Sharma, C. P. Zinc Is Critically Required for Pollen Function and Fertilisation in Lentil. *J. Trace Elem. Med. Biol.* **2006**, *20*:89–96.
85. Hu, Y.; Wu, S.; Sun, Y.; Li, T.; Zhang, X.; Chen, C.; Chen, B.; Arbuscular Mycorrhizal Symbiosis Can Mitigate the Negative Effects of Night Warming on Physiological Traits of *Medicago truncatula* L. *Mycorrhiza* **2015**, *25* (2), 131–142.
86. Fan, Y.; Lv, Z.; Qin, B.; Yang, J.; Ren, K.; Liu, Q.; Huang, Z. Night Warming at the Vegetative Stage Improves Pre-Anthesis Photosynthesis and Plant Productivity Involved in Grain Yield of Winter Wheat. *Plant Physiol. Biochem.* **2022**, *186*, 19–30.
87. Jagadish, S. V. K.; Murty, M. V. R.; Quick, W. P. Rice Responses to Rising Temperatures–Challenges, Perspectives and Future Directions. *Plant, Cell & Environment* **2015**, *38* (9), 1686–1698.

88. Sekizuka, F., Nose, A., Kawamitu, Y., Akinaga, T., Taira, C.; Onaha, A. Effect of Day/Night Temperature Conditions on CO_2 Exchange Rate and CO_2 Balance of Dendrobium Ekapol Panda No. 1. *Rec. Adv. Hortic. Sci. Tropics* **1990,** *292*, 187–194.
89. Lemoine, R.; Camera, S. L.; Atanassova, R.; Dédaldéchamp, F.; Allario, T.; Pourtau, N.; Bonnemain, J. L.; Laloi, M.; Coutos-Thévenot, P.; Maurousset, L.; Faucher, M. Source-to-Sink Transport of Sugar and Regulation by Environmental Factors. *Front. Plant Sci.* **2013**, 4, 272.
90. Glaubitz, U.; Li, X.; Schaedel, S.; Erban, A.; Sulpice, R.; Kopka, J.; Hincha, D. K.; Zuther, E. Integrated analysis of Rice Transcriptomic and Metabolomic Responses to Elevated Night Temperatures Identifies Sensitivity-and Tolerance-Related Profiles. *Plant Cell Environ.* **2017**, *40* (1), 121–137.
91. Seydel, C.; Kitashova, A.; Fürtauer, L.; Nägele, T. Temperature-Induced Dynamics of Plant Carbohydrate Metabolism. *Physiol. Plantar.* **2022**, *174* (1), e13602.
92. Schaarschmidt, S.; Glaubitz, U.; Erban, A.; Kopka, J.; Zuther, E. Differentiation of the High Night Temperature Response in Leaf Segments of Rice Cultivars with Contrasting Tolerance. *Int. J. Mol. Sci.* **2021**, *22* (19), 10451.
93. Chen, Y.; Wang, X. M.; Zhou, L.; He, Y.; Wang, D.; Qi, Y. H.; Jiang, D. A. Rubisco Activase Is Also a Multiple Responder to Abiotic Stresses in Rice. *PLoS one* **2015**, *10* (10), e0140934.
94. IPCC. Synthesis Report. In: *Contribution of Working Groups I, II and III to the Fifth Assessment Report of the Intergovernmental Panel on Climate Change*; Core Writing Team, Pachauri, R. K., Meyer, L. A., Eds.; IPCC: Geneva, 2014.
95. Hildebrandt, T. M.; Nesi, A. N.; Araujo, W. L.; Braun, H. P. Amino Acid Catabolism in Plants. *Mol. Plant* **2015**, *8*, 1563–1579.
96. Krasensky, J.; Jonak, C. Drought, Salt, and Temperature Stress-Induced Metabolic Rearrangements and Regulatory Networks. *J. Exp. Bot.* **2012**, *63*, 1593–1608.
97. Obata, T.; Fernie, A. R. The Use Metabolomics to Dissect Plant Responses to Abiotic Stresses. *Cell. Mol. Life Sci.* **2012**, *69*, 3225–3243.
98. Kumar, V.; Singhal, R. K.; Hidangmayum, A. L; Singh, A.; Aanjna, B.; Chauhan, J.; Bose, B. Anthropogenic Stress and Phenolic Compounds: An Environmental Robustness Diagnostics Compound Family in Stress Ameliorations. In: *Plant Phenolics in Abiotic Stress Management*; Springer Nature: Singapore, 2003; pp 391–413.

CHAPTER 5

Effect of Elevated Night Temperature on Nutrient Cycle, Nutrient Uptake, and Transport

EETELA SATHYANARAYANA[1], TEJA V. SURYA[2], S. SARANYA[3], J. BHARGHAVI[4], M. RAJASHEKHAR[5], RAJASHEKAR B[6], K. CHARAN TEJA[7], SAIDEEP THALLAPALLY[8], B. RAKESH[9], and VILAKAR KAYITHA[10]

[1]*Department of Soil Science and Agricultural Chemistry, Agricultural College, PJTSAU, Palem, India*

[2]*Department of Soil Science and Agricultural Chemistry, Assam Agricultural University, Jorhat, India*

[3]*GD Goenka University, Gurugram, Haryana, India*

[4]*Department of Crop Physiology, Agricultural College, PJTSAU, Warangal, India*

[5]*Agricultural Entomology, PJTSAU, Hyderabad, India*

[6]*KVK, PJTSAU, Palem, Hyderabad, India*

[7]*Department of Agronomy, Agricultural College, PJTSAU, Palem, India*

[8]*Forest College and Research Institute, Mulugu, Telangana, India*

[9]*Department of Soil Science and Agricultural Chemistry, TNAU, Coimbatore, India*

[10]*Keladi Shivappa Nayaka University of Agricultural and Horticultural Sciences (KSNUAHS), Shimoga, India*

Impact of High Night Temperature on Plant Biology: Toward Sustainable Plant Adaptation to Climate Change. Rajesh Kumar Singhal, Raju Bheemanahalli, Saurabh Pandey, and Pratibha M. D. (Eds.)

ABSTRACT

Temperature is the measure of heat. The rise or fall in temperatures will adversely affect the soil temperature. Soil temperature is a function of heat flux within the soil, as well as heat exchanges between the soil and atmosphere. An important source of soil temperature is radiation. Soil temperature varies seasonally and daily, which can result from changes in energy and energy changes going down through the soil surface. The temperature, especially at the far end of the night, has an impact on soil diurnal rhythms and affects the subsurface temperature. In the near future, an increase in soil temperature may cause accelerated rates of mineral weathering and decomposition, increasing the content of low-activity clays and decreasing organic matter. The quanta of radiation received by the soil affect soil temperature, biological processes like seed germination, seedling emergence, plant root growth, and therefore the availability of nutrients. Soil temperature also affects soil water retention, transmission, and availability to plants by inhibiting the diffusion and mass flow of nutrients.

5.1 INTRODUCTION

Soil acts as a major storehouse for heat that has an energy reservoir during the day and a heat source for the surface soil during the night. Solar energy is stored by the soil during the warm season, and it is released during the cold season.[1] The soil temperature is mostly dependent on the rate of energy absorbed to that lost from the soil. The fluctuations of soil temperature are annual and daily, which are affected significantly by variations in air temperature and solar radiation.[2] Soil temperature is one of the major factors that influence soil processes involved in plant growth. Soil temperature also influences inter-spheric processes such as the gaseous exchange between the atmosphere and soil.[3] The quantum of radiation impregnated into the soil also affects the soil temperature. The soil temperature changes the rate of decomposition of organic matter and the mineralization of different organic components in the soil.

Many years ago, Bowen[4] underlined the curious, indisputable fact that soil temperature had received comparatively less attention from scientists, even supposing it was of basic importance to plant productivity. Soil temperature is a function of heat flux in the soil and also of heat exchange between the soil and atmosphere.[5] It is also defined as the function of the internal energy of the soil. The important causes for the variations in soil temperature are the transfer of heat in the soil[6] and latent heat exchange at the surface of the soil with the atmosphere.[7] The cause of soil temperature is solar radiation. A thermometer is used for the measurement of soil temperature. The seasonal and daily variations in soil temperature are due to changes in radiation from the sun and energy transformations occurring through the soil surface.[8] Though many concluded that soil temperatures could rise due to various anthropogenic interventions, there is a renewed interest in learning the mechanisms and effects of soil temperature on terrestrial ecosystem responses.[9–11] However, this line of investigation continued to be underrepresented. In apprehensive consideration of soil temperature, we should tend to be cognizant that it regularly interacts with the organic phenomena and abiotic parts of the soil–plant system in both area and time.[12] For example, not only will soil temperature influence rate constants of chemical reactions, water content, and nutrient transport in soil, but it will also affect plant physiological aspects of nutrient uptake, root growth, and the composition and performance of soil microbial communities. Nearly all processes that occur in soil, right from the weathering of primary minerals to plant nutrition and the storage of organic carbon, are powerfully influenced by soil temperature. Above all, the acquisition of mineral nutrients by plants is ruled by myriad chemical, physical, and biological processes that are sensitive to temperature. Soil temperature thus exerts a significant influence on primary productivity in terrestrial ecosystems around the globe.

Root zone temperature (RZT) is one of the major factors influencing plant growth and nutrient and water uptake.[13–15] Various studies have significantly shown that plant growth is greatly affected by RZT.[16–18] Plant growth is influenced by numerous variables, of which root temperature and nutrient supply are considered to be important. Precisely, nutrient uptake is influenced by soil temperature.[19] Soil temperature affects the physicochemical and biological processes that influence nutrient availability in soils, which in turn affects the uptake of nutrients by plants.[20] The specific absorption rates of plant nutrients are highly dependent on the temperature of the soil, where even a slight increase in soil temperature might induce larger changes in plant growth and nutrient absorption, as reported by Clarkson et al.,[21] which signifies that RZT has a crucial role in the uptake of nutrients by the plant and their utilization.

5.2 EFFECT OF SOIL TEMPERATURE ON NUTRIENT CYCLE, NUTRIENT UPTAKE, AND NUTRIENT TRANSPORT

As such, soil temperature does not directly affect nutrient uptake and transport but indirectly affects them. Since nutrient cycling and nutrient uptake are interrelated with many other soil properties, the soil temperature alone cannot directly influence those factors. The various factors that are influenced by the soil temperature that affect nutrient uptake and transport are: bioactivity, soil microorganisms, organic matter decomposition, cation exchange capacity, soil pH, soil aeration, and water uptake (Fig. 5.1).

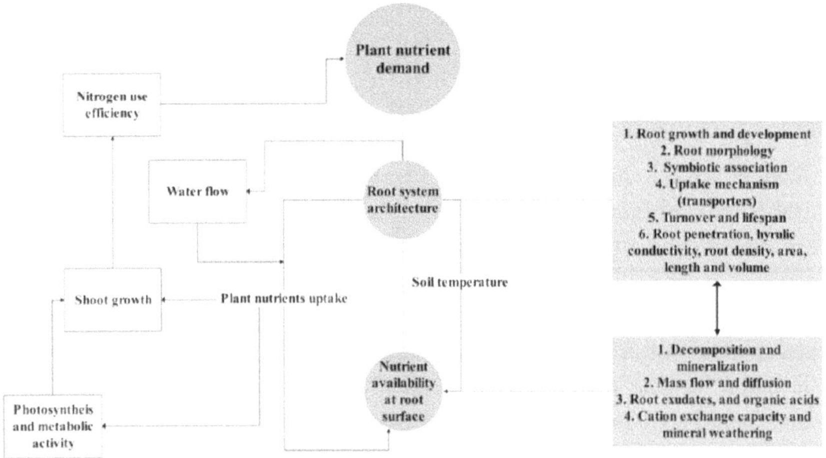

FIGURE 5.1 A conceptual diagram depicting relationships between plant and soil factors that control the availability and uptake of mineral nutrients. Soil temperature affects nutrient uptake directly by altering root growth, morphology, and uptake kinetics. Indirect effects include altered rates of decomposition and nutrient mineralization, mineral weathering, and nutrient transport processes.

Source: Reprinted/Adapted with permission from Ref. [41]. Copyright © 2005, Springer-Verlag Berlin Heidelberg

5.2.1 *BIOACTIVITY*

Soil temperatures that range between 10 and 28°C influence soil respiration by extravagating the activity of extracellular enzymes, which degrade organic matter content that contains polymeric substances in soils[22];

increasing the retake of soluble substrates by microbes[23]; and increasing the rate of respiration by microbes.[24] An increase in soil temperature enhances the mineralization of nitrogen through an increase in the activity of microbes and an increase in organic matter decomposition in the soil.[25] The mineralization of different nutrients decreases by inhibiting microbial activity and decreasing the diffusion of soluble substrates in the soil below the freezing point.[26]

5.2.2 SOIL MICROORGANISMS

The activity of most of the soil microorganisms occurs in temperature ranges of 10–35.6°C. Most of the activities decrease at low soil temperatures and cease at the freezing point.[27] Night warming increases microbial respiration and thereby increases the availability of various nutrients essential for plant growth.

5.2.3 ORGANIC MATTER DECOMPOSITION

The accumulation of soil matter increases at temperatures below 0°C due to the slower pace of decomposition rate.[23] At lower temperatures, there will be a decrease in microbial activity and biochemical processes, which decrease the decomposition of organic matter.[28] Decomposition of organic matter increases in the soil temperature range between 2 and 38°C[29] by enhancing the mobility of soluble substrates in the soil [30] and stimulating the activities of microbes.[31]

5.2.3.1 EFFECT OF ELEVATED NIGHT TEMPERATURE ON CARBON AND NITROGEN ASSIMILATION

Carbon assimilation decreases with an increase in temperature during the nocturnal hours, whereas the assimilation of nitrogen increases with an increase in night temperature (Table 5.1). However, the total biomass content increases with elevated night temperatures due to an increase in night respiration.[32]

TABLE 5.1 Effects of Elevated Night Temperature on the C and N Concentrations and the Leaf Area per Unit Weight.

Night temperature	C concentration		N concentration		Leaf area per weight ($cm^2\ gm^{-1}$)
	Living leaf (%)	Dead leaf (%)	Living leaf (%)	Dead leaf (%)	
High (32° C)	43.8	41.1	1.14	0.62	159.6
Low (22° C)	44.1	41.9	1.03	0.61	138.2

Source: Reprinted/Adapted with permission from Ref. [33]. Copyright © 2010 Elsevier B.V.

5.2.4 CATION EXCHANGE CAPACITY

An increase in the temperature of the soil decreases organic matter content[34] by combustion. This decrease in the organic matter content and reduction in the clay size fraction of soil due to elevated temperatures leads to a decrease in the cation exchange capacity of soil.[35] An increase in the soil temperature leads to a decrease in the cation exchange capacity of the soil, as reported by Rengasamy and Churchman.[36]

5.2.5 SOIL PH

At a soil temperature range between 25 and 39°C, the pH of the soil increases due to organic acid denaturation, and it increases with an increase in the temperature.[37]

5.2.6 SOIL AERATION

The percentage of carbon dioxide content is influenced by the soil temperature. Higher temperatures stimulate the activity of microbes, which in turn results in a higher production of carbon dioxide in the soil by microbes.[27]

5.2.7 WATER UPTAKE

At low temperatures, there will be a decrease in the uptake of water. At low temperatures, the viscosity of the solution is increased and the absorption of the water is decreased,[38] associated with a decrease in the rate of photosynthesis.

5.2.8 AVAILABLE PHOSPHOROUS

Yilvaiaio et al.[39] reported that there is an increase in water-soluble phosphorus between the soil temperature range of 5 and 25°C due to the increase in the mobility of phosphorus in the soil that is controlled by diffusion. The availability of phosphorous in soils decreases with a decrease in soil temperature because the mineralization of phosphorus from organic matter ceases at lower temperatures.[40,42]

5.3 NIGHT-TIME WARMING AFFECTS N AND P DYNAMICS

The high night temperatures (HNT) generated by global warming have been shown in a number of studies to have major effects on the grain yield and biomass production of major agricultural crops. For example, Fang et al.[43] found that winter wheat yields dropped by 27% when the night-time temperature was elevated by 2.5°C. This was due to the fact that HNTs caused a considerable increase in the number of sterile tillers, fewer spikelets, and fewer grains per spike. The wheat output will decrease as a result of a shorter growth period, as anticipated by the CERES model, if the daily minimum temperature continues to climb.[44] Recent studies have shown, however, that rises in night-time temperatures have had minimal impact on winter wheat yield in northern China.[45] The findings of Zheng et al.[46] revealed that an increase in night-time temperature led to an increase in the yield potential of winter wheat due to a decrease in preanthesis time and an increase in postanthesis time. On the other hand, an increase in night-time temperatures (NTs) may lead to an increase in nitrogen (N) and phosphorus (P) soil mineralization, and as a result, a change in the amounts of N and P that are absorbed by plant roots.[47] In addition, the changed physiological processes that occurred as a result of the increased temperature had a substantial impact on the plant's internal nutrient intake,[48] which had a considerable impact on the plant's overall production. There have been a few studies that have established the effect that an increase in temperature has on the absorption and allocation of nitrogen in a few different plant species. Despite the fact that winter wheat is typically subjected to a greater negative influence of night-time warming during the winter and early spring, no study has simultaneously investigated the effects of HNTs on the accumulation and translocation of nitrogen and phosphorus in wheat. This is despite the fact that high night-time temperatures have the potential to affect the accumulation of nitrogen and phosphorus.[49]

The intake of nutrients (N and P) is one of the most critical factors determining the photosynthetic capability and plant output. Numerous studies have demonstrated that heat may either increase or decrease N accumulation in the plants under observation. The majority of these studies were conducted in natural ecosystems without N fertilizer. For instance, experimental warming over a 4-year period boosts grass N accumulation by 5–20% because increases in above-ground biomass are larger than decreases in leaf N concentrations.[50] Other studies conducted in growth chambers revealed that HNTs decreased above-ground biomass and nitrogen concentration in the leaves, which in turn decreased plant N accumulation.[51] Thus, the combined responses of plant biomass and N concentration to warming are what significantly drive plant N accumulation at HNTs.[50] High night-time temperatures had little influence on whole plant N concentration, according to a recent study using rice crops grown in controlled-environment chambers, but they significantly enhanced N absorption via increased above-ground plant biomass.[52] Regarding P, prior studies have shown that higher temperatures, even those at night, may enhance soil phosphatase activity and so accelerate soil P mineralization,[53] leading to a 42% increase in P content in plant leaves.[54] However, some research has shown that a reduction in soil moisture brought on by a rise in temperature may also decrease the capacity of the soil to diffuse nutrients and the activity of certain soil and plant root enzymes.[55] Furthermore, since microorganisms in soil fight for P more fiercely than plants do, the rise in available P in soil brought on by warming has not been matched by a rise in P accumulation in above-ground biomass.[56]

Wheat production may be significantly impacted by a number of other plant variables, such as N and P translocations.[57] Field-scale warming modulation may decrease N translocation from senescent to surviving leaves in grass or boost P allocation from stem to leaves in shrubs. A recent study by Cheng et al.[52] found that a high night-time temperature prevented nitrogen (N) from moving from vegetative parts to rice ears during reproductive development, which reduced grain yield. The physiological responses of N and P in winter wheat plants to high night-time temperatures and their potential impacts on grain yield are still poorly understood.

5.4 CONCLUSIONS

Soil temperature features a strong influence on the community composition and activity of the soil fauna and microflora, affecting rates of organic matter decomposition and nutrient mineralization. The abundance and activity of soil fauna generally increase with rising soil temperatures, especially in cold soils, but are often limited by the drying of the soil at high temperatures. A key area of future research is to work out how rising soil temperature affects community composition and substrate utilization of soil fauna and microflora communities, and therefore the effects on nutrient availability. The symbiotic organic process appears to increase with rising soil temperatures, especially at low temperatures, thanks to increasing nodule mass and activity. However, laboratory studies indicate that rates of inorganic N uptake may increase to a greater extent, dominating total N nutrition. A key area of future research is to work out to what extent symbiotic N fixation is stimulated by rising soil temperatures in the field, for systems during which this is an important part of overall N nutrition. Growth rates and community composition of mycorrhizae have also been shown to be sensitive to changes in soil temperature, but how these changes affect the acquisition of nutrients and C cycling in forest ecosystems isn't well understood. A key area of future research is to work out the extent to which mycorrhizal nutrient uptake and turnover are influenced by rising soil temperature in natural ecosystems. In summary, the soil temperature is one of the key factors regulating nutrient availability and uptake in terrestrial ecosystems. Changes in soil temperature are inextricably linked to changes in soil moisture at the ecosystem level. Changes in plant and microbial function are going to be very tightly coupled to changes in soil temperature and moisture in the future, and these ecological interactions will play critical roles in nutrient availability and net ecosystem productivity.

KEYWORDS

- **soil temperature**
- **diurnal rhythms**
- **nutrient uptake**
- **nutrient cycling**
- **nutrient transport**

REFERENCES

1. Geiger, R.; Aron, R. N.; Todhunter, P. *The Climate Near the Ground*; Rownaan and Little Field Publishers, Inc.: Lanham, 2003; pp 42–50.
2. Wu, J.; Nofziger, D. L. Incorporating Temperature Effects on Pesticide Dug Radiation into a Management Model. *J. Environ. Qual.* **1999,** *28*, 92–100.
3. Lehnert, M. The Soil Temperature Regime in the Urban and Suburban Landscapes of Olomoric, Czech Republic. *Morarian Geograph. Rep.* **2013,** *21* (3), 27–36.
4. Bowen, G. D. Soil temperature, Root Growth, and Plant Function. In: *Plant Roots: The Hidden Half*; Waisel, Y., Eshel, A., Kafkafi, U., Eds.; Marcel Dekker: New York, 1991; pp 309–330.
5. Elias, E. A.; Cichota, R.; Torraiani, H. H. Analytical Soil Temperature Model: Correction for Temporal Variation of Daily Amplitude. *Soil Sci. Soc. Am. J.* **2004,** *68* (3), 784–788.
6. Zhao, Y. M.; Xao Mong, M. A.; Wang, L. Variations of soil temperature and Soil Moisture in Northern Tibetaplateau. *J. Glaciol. Georcyol.* **2007,** *29* (4), 578–583.
7. Nwankwo, C.; Ogugurue, D. An Investigation of Temperature Variation at Soil Depths in Peuts of Southern Nigeria. *Am. J. Environ. Eng.* **2012,** *2* (5), 142–147.
8. Chiemeka, I. U. Soil Temperature Profile at Uturu, Nigeria. *Pac. J. Sci. Technol.* **2010,** *11* (1), 478–482.
9. Billings, W. D.; Luken, O. J.; Mortensen, D. A.; Peterson, K. M. Arctic Tundra: A Source or Sink for Atmospheric Carbon Dioxide in a Changing Environment? *Oecologia* **1982,** *53*, 7–11.
10. Peterjohn, W. T.; Melillo, J. M.; Steudler, P. A.; Newkirk, K. M.; Bowles, F. P.; Aber, J. D. Responses of Trace Gas Fluxes and N Availability to Experimentally Elevated Soil Temperatures. *Ecol. Appln.* **1994,** *4*, 617–625.
11. Rosenzweig, C.; Hillel, D. Soils and Global Climate Change: Challenges and Opportunities. *Soil Sci.* **2000,** *165*, 47–56.
12. Pregitzer, K. S.; King, J. S.; Burton, A. J.; Brown, S. E. Responses of Tree Fine Roots to Temperature. *New Phytol.* **2000,** *147*, 105–115.
13. Mozafar, A.; Schreiber, P.; Oertli, J. J. Photoperiod and Root Zone Temperature: Interacting Effects on Growth and Mineral Nutrients of Maize. *Plant Soil* **1993,** *153*, 71–78.
14. Marschner, H.; Kirkby, E. A.; Cakmak, I. Effect of Mineral Nutritional Status on Shoot-Root Partitioning of Photo Assimilates and Cycling of Mineral Nutrients. *J. Exp. Bot.* **1996,** *47*, 1255–1263.
15. Bode Stoltzfus, R. M.; Taber, H. G.; Aiello, A. S. Effect of Increasing Root-Zone Temperature on Growth and Nutrient Uptake by 'Gold Star' Muskmelon. *J. Plant Nutri.* **1998,** *21*, 321–328.
16. Solfjeld, I.; Johnsen. The Influence of Root-Zone Temperature on Growth of *Betula pendula* Roth. *Trees* **2006**, *20*, 320–328.
17. Dı´az-Pe´rez, J. C.; Gitaitis, R.; Mandal, B. Effects of Plastic Mulches on Root Zone Temperature and on the Manifestation of Tomato Spotted Wilt Symptoms and Yield of Tomato. *Sci. Hortic.***2007,** *114*, 90–95.

18. Nxawe, S.; Laubscher, C. P.; Ndakidemi, P. A. Effect of Regulated Irrigation Water Temperature on Hydroponics Production of Spinach (*Spinacia oleracea* L). *Afr. J. Agric. Res.* **2009,** *4,* 1442–1446.
19. Xu, Q.; Huang, B. Seasonal Changes in Root Metabolic Activity and Nitrogen Uptake for Two Cultivars of Creeping Bent Grass. *Hortic. Sci.* 2006, *41*, 822–826.
20. Hussain, A.; Maqsood, M. A. Root Zone Temperature Influences Nutrient Accumulation and Use in Maize. *Pak. J. Bot.* **2011,** *43,* 1551–1556.
21. Clarkson, D. T.; Jones, L. H. P.; Purves, J. V. Absorption of Nitrate and Ammonium Ions by *Lolium perenne* from Flowing Solution Cultures at Low Root Temperatures. *Plant Cell Environ.* **1992,** *15,* 99–106.
22. Conant, R. T.; Drijber, R. A.; Haddix, M. C. Sensitivity of Organic Matter Decomposition to Warming Varies With Its Quality. *Glob. Chang. Biol.* **2008,** *14* (4), 868–877.
23. Allison, D.; Wallenstein, M. D.; Bradford, M. A. Soil Carbon Response to Warming Dependent on Microbial Physiology. *Nat. Geosci.* **2010,** *3,* 336–340.
24. Wallenstein, M. D.; Allisons, S.; Ernakovich, J. Controls on the Temperature Sensitivity of Soil Enzymes, a Key Driver of In Situ Enzyme Activity Rates. *Soil Enzymol.* **2010,** *15,* 245–258.
25. Yan, L.; Hangwen, X. Effects of Soil Temperature, Flooding and Organic Matter Addition in N_2O Emissions from a Soil of Hongze Lake Wetland, China. *J. Appl. Soil Ecol.* **2014,** *29,* 173–183.
26. Kaiser, C.; Meger, A.; Biasi, C. Conservation of Soil Organic Matter Through Crytoturbation in Artic Soils in Siberia. *J. Appl. Soil Ecol.* **2007,** *112,* 173–183.
27. Allison, S. O. Cheaters, Diffusion and Nutrient Content Decomposition by Microbial Enzymes in Spatially Structured Environments. *Ecol. Lett.* **2005,** *8* (6), 626–635.
28. Gilichinsky, D. A.; Rivikina, E. M. Permafrost Microbiology. In: *Encyclopaedia of Geo Biology*; Ratner, J., et al., Eds.; Springer: Netherlands, 2011; 726–732.
29. Broadbent, F. E.; Soil Organic Matter. *Sustain. Opt. Land Manag.* **2015,** *2,* 34–38.
30. Fierrer, N.; Gruine, J. M.; McLauchlan, K. Litter Quality and the Temperature Sensitivity of Decomposition. *Ecol.* **2005,** *86* (2), 320–326.
31. Fang, C. M.; Smith, P.; Monorieff, J. B. Similar Response of Labile and Resistant Soil Organic Matter Pools to Changes in Temperature. *Nature* **2005,** *436,* 881–883.
32. Cheng, W.; Sakai, H.; Yagi, K.; Hasegawa, T. Interactions of Elevated [CO_2] and Night Temperature on Rice Growth and Yield. *Agric. For. Meteorol.* **2009,** *149,* 51–58.
33. Cheng, W.; Sakai, H.; Yagi, K.; Hasegawa, T. Combined Effects of Elevated [CO_2] and Night Temperature on Carbon Assimilation, Nitrogen Absorption, and the Allocations of C and N by Rice (*Oryza sativa* L.). *Agric. For. Meteorol.* **2010,** *150,* 1174–1181.
34. Ubeda, X.; Pereira, P.; Outeiro, L.; Effects of Fire Temperature on the Physical and Chemical Characteristics of the Ash from Two Plot of Cork Oak (*Quercus suber*). *Land Degrad. Dev.* **2009,** *20* (6), 589–608.
35. Certini, G. Effects of Fire in Properties of Forest Soils: A Review. *Oecology* **2005,** *143* (1), 1–10.

36. Rengasmy, P.; Churchman, G. J.; Cation echange capacity, exchangeable cations and Solidity. In: *Soil Analysis*; Peverrilli, K. I. et al., Eds.; An Interpretation Manual CSIRO Publishing: Australia, 2009.
37. Mensies, N.; Gillman, G. Plant Growth Limitation and Nutrient Loss Following Piled Burning in Slash and Burn Agriculture. *Nutr. Cycl. Agroecosys.* **2003,** *65* (1), 23–33.
38. Toselli, M.; Flore, J. A.; Marogoni, B. Effects of Root–Zone Temperature on Nitrogen Accumulation by Non–Breeding Apple Trees. *J. Hortic. Sci. Biotechnol.* **1999,** *74*, 118–124.
39. Yilvainio, K.; Pettovuori, T. Phosphorus Acquisition by Barley (*Hordeum vulgare*) at Suboptimal Soil Temperature. *Agric. Food Sci.* **2012,** *21*, 453–461.
40. Gahoonia, T. S.; Nielsen, N. E. Phosphorus Uptake and Growth of Root Hairless Barley Mutant (Bald Root Barley) and Wild Type in Low and High–P Soils. *Plant Cell Environ.* **2003,** *26*, 1759–1766.
41. Pregitzer, K. S.; King, J. S. Effects of Soil Temperature on Nutrient Uptake. Ecological Studies. In: *Nutrient Acquisition by Plants: An Ecological Perspective*; BassiriRad, H., Ed., Vol. 181; Springer-Verlag: Berlin Heidelberg, 2005; pp 277–310.
42. Elbasiouny, H.; El-Ramady, H.; Elbehiry, F.; Rajput, V.D.; Minkina, T.; Mandzhieva, S. Plant Nutrition Under Climate Change and Soil Carbon Sequestration. *Sustainability* **2022,** *14* (2), 1–20.
43. Fang, S.; Tan, K.; Ren, S. Winter Wheat Yields Decline with Spring Higher Night Temperature by Controlled Experiments. *Sci. Agric. Sinica* **2010,** *43* (15), 3251–3258.
44. Wilkens, P.; Singh, U. A Code-Level Analysis for Temperature Effects in the CERES Models. In: *Modeling Temperature Response in Wheat and Maize*; 2003; p 1.
45. Fang, S.; Tan, K.; Ren, S.; Zhang, X.; Zhao, J. Fields Experiments in North China Show no Decrease in Winter Wheat Yields with Night Temperature Increased by 2.0–2.5 C. *Sci. China Earth Sci.* **2012,** *55*, 1021–1027.
46. Zheng, C.; Zhang, J.; Chen, J.; Chen, C.; Tian, Y.; Deng, A.; Song, Z.; Nawaz, M. M.; van Groenigen, K. J.; Zhang, W. Night-time Warming Increases Winter-Sown Wheat Yield Across Major Chinese Cropping Regions. *Field Crops Res.* **2017,** *214*: 202–210.
47. Mohammed, A. R.; Tarpley, L. Effects of High Night Temperature and Spikelet Position on Yield-Related Parameters of Rice (*Oryza sativa* L.) Plants. *Eur. J. Agron.* **2010,** *33* (2), 117–123.
48. Zhang, Y.; Li, R.; Wang, Y. Night-Time Warming Affects N and P Dynamics and Productivity of Winter Wheat Plants. *Can. J. Plant Sci.* **2013,** *93* (3), 397–406.
49. Beier, C.; Emmett, B. A.; Peñuelas, J.; Schmidt, I. K.; Tietema, A.; Estiarte, M.; Gundersen, P.; Llorens, L.; Riis-Nielsen, T.; Sowerby, A.; Gorissen, A. Carbon and Nitrogen Cycles in European Ecosystems Respond Differently to Global Warming. *Sci. Total Environ.* **2008,** *407* (1), 692–697.
50. Xia, J.; Niu, S.; Wan, S. Response of Ecosystem Carbon Exchange to Warming and Nitrogen Addition During Two Hydrologically Contrasting Growing Seasons in a Temperate Steppe. *Glob. Chang. Biol.* **2009,** *15* (6), 1544–1556.
51. Xu, Z. Z.; Zhou, G. S. Effects of Water Stress and High Nocturnal Temperature on Photosynthesis and Nitrogen Level of a Perennial Grass *Leymus chinensis*. *Plant Soil* **2005,** *269*: 131–139.

52. Cheng, W.; Sakai, H.; Yagi, K.; Hasegawa, T. Combined Effects of Elevated [CO_2] and High Night Temperature on Carbon Assimilation, Nitrogen Absorption, and the Allocations of C and N by Rice (*Oryza sativa* L.). *Agric. For. Meteor.* **2010,** *150* (9), 1174–1181.
53. Sardans, J.; Peñuelas, J.; Estiarte, M. Warming and Drought Alter Soil Phosphatase Activity and Soil P Availability in a Mediterranean Shrubland. *Plant Soil* **2006,** *289*, 227–238.
54. Sardans, J.; Peñuelas, J.; Prieto, P.; Estiarte, M. Drought and Warming-Induced Changes in P and K Concentration and Accumulation in Plant Biomass and Soil in a Mediterranean Shrubland. *Plant Soil* **2008,** *306*, 261–271.
55. Llorens, L.; Peñuelas, J.; Estiarte, M.; Bruna, P. Contrasting Growth Changes in Two Dominant Species of a Mediterranean Shrubland Submitted to Experimental Drought and Warming. *Ann. Bot.* **2004,** *94* (6), 843–853.
56. Kellogg, L. E.; Bridgham, S. D. Plant-Microbe Competition for Phosphorus in Three Vertical Location. *Ecol. Soc. Am. Ann. Meet. Abs.* **2001,** *86*, 128.
57. Dordas, C. Dry Matter, Nitrogen and Phosphorus Accumulation, Partitioning and Remobilization as Affected by N and P Fertilization and Source–Sink Relations. *Eur. J. Agron.* **2009,** *30* (2), 129–139.

CHAPTER 6

Effect of Elevated Night Temperature on Metabolism (Respiration)

DASARI SREEKANTH[1], PAWAR DEEPAK VISHWANATH[1], M. NAGARAJU[5], RAJEEV KUMAR[2], SHOBHA SONDHIA[1], P. K. SINGH[1], ASHISH MARATHE[3], RATNA KUMAR PASALA[6], B. KIRAN KUMAR[5], and SUBHASH CHANDER[4]

[1]*ICAR-Directorate of Weed Research, Jabalpur, Madhya Pradesh, India*

[2]*ICAR-Indian Institute of Vegetable Research, Varanasi, Uttar Pradesh, India*

[3]*ICAR-National Institute of Biotic Stress Management, Raipur, Chhattisgarh, India*

[4]*ICAR-National Bureau of Plant Genetic Resources, New Delhi, India*

[5]*Department of Botany, Osmania University, Hyderabad, Telangana, India*

[6]*ICAR-Indian Institute of Oilseeds Research, Hyderabad, Telangana, India*

ABSTRACT

Under the projected climate, high night-time temperatures (HNTs) are a significant obstacle to maintaining global food security. By the end of this century, global climate models indicate that the average temperature will have increased by 0.3–4.8°C. Various authors have documented strong evidence that minimum night-time temperatures (NTs) rise far more quickly than maximum daytime temperatures. Studies on the effects of

Impact of High Night Temperature on Plant Biology: Toward Sustainable Plant Adaptation to Climate Change. Rajesh Kumar Singhal, Raju Bheemanahalli, Saurabh Pandey, and Pratibha M. D. (Eds.)

high daytime temperatures (DTs) and NTs on plant carbon relations are necessary due to predicted patterns of global warming, the differences in how respiration and photosynthesis react to temperature, and the relationship between these two processes. According to many investigations, HNT causes higher respiration that leads to carbon loss (night respiration), decreases grain and total dry weight, reduces photosynthesis, increases membrane damage, and, in the end, reduces fertility and yield. Plants with increasing temperatures have increased respiratory activity that is thermodynamically driven, which may be the cause of 30–80% of the loss of absorbed carbon. On the contrary, recent research suggests that HNT can influence the translocation and accumulation of photosynthates during the grain-filling process. Therefore, it is important to investigate how HNT affects carbon balance and the activity of key source and sink enzymes that significantly control the generation and storage of photoassimilates in the source and sink tissues during the crucial grain-filling phase. To address the impacts of HNT on plant metabolism (respiration), future work is urgently required to understand the complex mechanisms under HNT to mitigate the adverse effects of climate change.

6.1 INTRODUCTION

In recent decades, the growing intensity of human activity has magnified climate change and its implications.[1] Global warming is a notable aspect of climate change, induced by man-made emissions of important greenhouse gases such as CO_2, CH_4, N_2O, deforestation, and urbanization. The global surface temperature has risen by 0.85°C on average over the last century, with a further rise of up to 3.7°C expected by 2100.[2] This rise in atmospheric temperature is driven by an upsurge in the average daily minimum temperature, rising faster than the average daily maximum temperature.[3] However, recent climate projections indicated that night-time temperatures (NTs) are expected to upsurge more quickly than daytime temperatures (DTs). This temperature increase occurs asymmetrically, with a more remarkable rise in daily minimum temperatures compared to daily maximum temperatures, resulting in a 'high night temperature' (HNT).[3]

Heat stress is most vulnerable at almost all stages of growth and development, and the reproductive stage appears more vulnerable than the vegetative stage.[4] High and low-temperature extremes typically contribute to physiological disorders and dysfunction, resulting in yield loss and quality degradation.[5] NT is one of the key environmental elements impacting plant metabolic activities.[6] HNT causes increased carbon loss from night respiration and decreases the activity of source-sink enzymes, resulting in lower biomass, poor grain filling, and lower grain weight.[7,8] Rice output and grain quality were significantly reduced by HNT at the farm and regional levels,[9] resulting in significant economic losses.[10]

HNT increases respiration and lowers the yield.[11] HNT increases respiration and its impact on biomass loss has been assessed to be up to 43 g/m^2, accounting for less than 20% of the decline.[12] A 1°C enhancement in NT can diminish rice grain production by 10%.[6] Elevated respiration caused by high temperatures can generate oxidative stress markers in various crops.[13] In general, enzymic and non-enzymic antioxidants reduce the harmful effects of oxidative stress markers; still, under stress, oxidants can overcome antioxidant levels, resulting in cell damage.[14] Reactive oxygen species (ROS)-induced cell damage can reduce cell membrane integrity, disturbing water, ion, and organic solute flow through the cell membrane and influencing carbon generation, consumption, passage, and buildup.[15]

Several studies revealed that HNT affected the reproductive development of numerous crops, such as rice and soybean, which affected the yield; however, the outcomes were inconsistent.[6] Gibson and Mullen[16] reported that HNT increased pod and seed numbers while decreasing seed size, not affecting seed yield. Furthermore, HNT induced flower opening and pod setting on secondary and tertiary racemes, increasing pod and seed numbers but reducing seed size, subsequently in an unaltered yield and harvest index.[11] Grain yield is influenced by carbon and nitrogen delivery to the grain and modulated by HNT.[17] Furthermore, HNT reduces panicle starch content and adversely influences grain yield and quality. HNT-induced physiological effects include increased respiration in leaves and panicles[7]; however, photosynthesis is unaffected or may be reduced. A drop in nitrogen and carbohydrate transportation after flowering was also explored as a probable reason for the yield decrease in HNT-sensitive cultivars.

Therefore, the present chapter deals with developments in understanding the impact of HNT on crop response and underlying physiological and biochemical mechanisms in plants.

6.2 EFFECT ON CARBON BALANCE AND ITS IMPACT ON YIELD

High temperatures affect both overall respiration and photosynthetic processes. It is generally observed that with an elevation in leaf temperature, the net photosynthesis rises, peaks at an ideal temperature, and then starts to decrease. The patterns of photosynthesis (net CO_2 assimilation) and nocturnal respiration, which are crucial for maintaining the carbon balance in plants, could be remarkably changed by the local climate and the stage of development. The plant carbon equilibrium is established by the plant's capacity to absorb carbon during photosynthesis and expel the absorbed carbon during respiration, typically at night-time. Approximately 70% of the total carbon fixed during photosynthesis is lost during respiration.[18] In their study of the fluctuation in respiration, Swain et al.[19] found that rice leaf respiration was lowest at noon and highest at dawn and sunset, implying higher respiratory carbon emissions at night-time. Additionally, premature to mid-grain filling corresponds to maturity and a more extensive range of temperatures that wheat undergoes under the current growth circumstances, increasing the amount of carbon consumed during night-time respiratory processes.[20] Assimilates that might normally promote grain development are broken down more quickly due to enhanced respiratory processes.

Previous studies have reported that HNT impacts the yields in various crops, such as soybean and rice, by affecting their reproductive growth. However, the results are inconsistent. According to Morita et al.,[21] as the NT rose from 22 to 34°C, rice production declined by 11%; however, whenever NT raised from 22 to 32°C, rice production increased by 12%.[22] Djanaguiraman et al.[23] observed that when the NT rose from 20 to 29°C pod setting, pollination dramatically reduced in soybean. On the contrary, Zheng et al.,[11] at NT 20–30°C, observed that the pod formation and seed quantity rose; however, the seed size dropped. Several investigations have examined the impact of HNT on wheat production and noticed that yield drop in HNT is primarily determined by shorter critical growth periods, increased night-time respiration, lowered leaf photosynthesis, fewer antioxidant activities, less biomass accumulation, fewer grains, and seed setting.[24] Despite attempts to comprehend various physiologically, developmental, and yield-related aspects of wheat that HNT alters, underlying physiological modifications that cause HNT-induced responses have still not been discovered. Metabolite profiling is an essential approach for identifying heat stress-related changes in biosynthetic processes, which also identifies compounds that confer endurance and adaptation response and metabolites of stress physiology and cell signaling.[25] The capacity of floral organs to withstand or be susceptible to drought and heat strain was discovered to be controlled by carbohydrate metabolism in *Oryza sativa*.[26]

6.3 EFFECT ON POSTFLOWERING NIGHT RESPIRATION

Climate forecasts indicated NT is estimated to upsurge more rapidly than DT.[27] A significant determinant in plant development is temperature, with higher optimal temperatures interrupting physiological and metabolic processes and increasing output variability.[28] Carbon loss was substantially higher in the postflowering phase of rice than in the preflowering stage by HNT-induced enhancement in night respiration.[7]

The level of acclimatization to night respiration depends on the length of HNT exposure,[29] and the thermal adaptation of night respiration is unaffected by the diel cycle.[30] Cereals such as wheat and rice accumulate considerably less biomass after postflowering HNT treatment.[29] In winter wheat, a substantial enhancement in night respiration was observed after 4 days of HNT exposure but not after 8 days, indicating adaptation after prolonged exposure.[29] This is due to an increased need for carbon from enhanced night respiration under HNT exposure at the vegetative phase, which may be supplied by efficient photosynthesis.[31] or via reprogramming metabolites from the tricarboxylic acid (TCA) cycle.[30]

Numerous investigations have demonstrated the adverse impact HNT had on numerous crops, that is, rice (*O. sativa* L.),[32] cowpea (*Vigna unguiculata* L.),[33,34] wheat *Triticum aestivum*),[35] sorghum (*Sorghum bicolor* L.),[36] and soybean (*Glycine max* L.) has specified that HNT during the reproductive phase has a profound impact on yield.[23] The alternations induced in crops by HNT (Table 1).

TABLE 6.1 Impact of HNT on Various Physiological and Biochemical Indices.

S. No.	Alterations in crop plants by HNT	References
1	Decreasing total dry biomass and grain weight	[21]
2	Decreasing photosynthesis	[37]
3	Increased membrane damage	[32]
4	Decreasing spikelet fertility and yield	[17]
5	Enhanced male sterility and floral abscission	[38]
6	Floral bud suppression and decreased pollen viability	[17, 28]
7	Lower seed-set rate	[17, 35, 39]
8	Decreased boll setting	[40]
9	Increased fruit-shedding rates	[41]

Increased whole-plant senescence brought by postflowering HNT treatment limited resource transfer from source to sink because of the short grain-filling period.[42] Although rice subjected to HNT experienced an acceleration of the grain-filling rate at an early phase of grain formation (i.e., 1–10 days after anthesis), the rate decreased at the late stages of grain formation (i.e., 20 days after anthesis) compared to control, ultimately resulted in lesser grain weight under HNT.[43] This shows that a higher grain filling rate does not make up for a shorter time frame for filling grains. Similarly, the HNT-tolerant rice genotype N22 recorded a noticeably high initial maximum and mean grain-filling rate throughout the panicle. In contrast, the HNT-sensitive genotype Gharib recorded a higher initial grain-filling rate, but the maximum and mean grain-filling rates significantly decreased, lowering grain weight.[44] Lower yield in rice and wheat under HNT stress during grain-filling is due to a lack of assimilates to keep up with the enriched grain-filling and shorter grain-filling period.[35] Additionally, a meta-analysis of C_3, C_4, and Crassulacean acid metabolism (CAM) plants that examined the effects of HNT on plant growth and productivity identified a drop in assimilate partition to reproductive structures as the primary contributory factor to the lower yield.[45] Different studies have documented that HNT-induced various crop alterations are represented in Figure 6.1.

6.4 PARTITIONING DURING THE CRITICAL FLOWERING AND POSTFLOWERING GROWTH STAGES

HNT reduced assimilate allocation to reproductive organs was the primary factor contributing to yield drop in C_3, C_4, and CAM plants.[45] Reduced boll setting,[40] higher fruit shedding,[41] and a lessening in the seed number[46] have all been linked to lower yields in cotton. However, little attention has been paid to the impact of HNT on cotton's reproductive unit's physiology and metabolism. Hence, this needs to be taken care of in future studies.

Mohammed and Tarpley[47] reported that HNT changed pollen germination and spikelet fertility and impacted rice productivity. It was observed that yield loss in field-grown rice was noticed by HNT through various physiological processes, including a decreased carbohydrate pool, increased biomass, and higher night-time respiration compared to increased spikelet sterility under increased DT stress.[48] Similar to this,

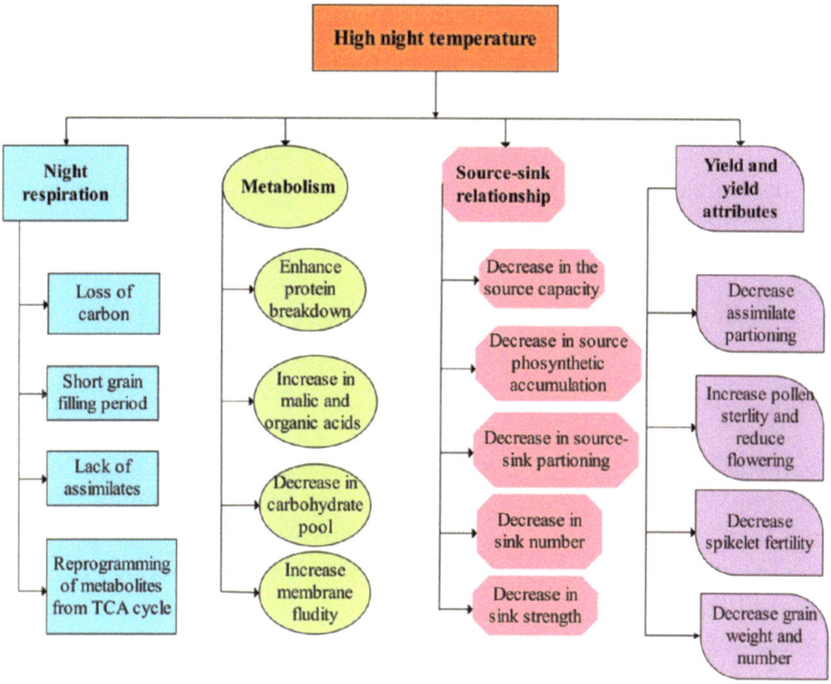

FIGURE 6.1 HNT-induced alterations in night respiration, metabolism, source–sink relationship, and yield and yield attributes.

HNT stress at flowering was primarily responsible for a drop in the spikes number in wheat. In contrast, HNT at grain filling caused a shortening of the grain-filling time, resulting in considerable weight loss.[42] This shows that HNT adversely influences wheat agronomic performance, indicating the necessity for a more thorough examination of the mechanisms underlying yield and quality losses under HNT exposure. HNT altered the carbon balance of the entire plant, increased night-time respiration, starch, and sucrose metabolism in wheat spikes, and metabolites connected to the TCA cycle in leaves.[29]

Increased respiration rates induce oxidative stress marker generation.[13] Depending on the free radical scavenging activity of the crop, oxidative stress markers can have toxic effects on plant metabolism. According to Frottin et al.,[49] glutathione reductase (GR) is a vital player in the construction of meristems and embryos, and its levels in the anthers regulate male sterility.[50]

Due to the reduced grain-filling period caused by HNT exposure postflowering, the transfer of resources from source to sink was constrained.[35,42,51] Although rice under HNT experienced an acceleration of the grain-filling rate during the early stages of grain development (e.g., 1 to 10 days after anthesis), the rate decreased during the late stages (e.g., 20 days after anthesis), ultimately leading to declining in grain weight at HNT.[43]

6.5 ON SINK-STRENGTH RELATED ENZYMATIC ACTIVITY DURING PEAK GRAIN FILLING PHASE AFFECTING RICE GRAIN YIELD AND QUALITY

Temperature and other abiotic stresses slowed the grain filling rate by reducing enzyme activity, further reducing starch biosynthesis.[52] The source–sink relationship in plants, particularly the grain-filling stage, is vital for understanding the grain-filling problem. Grain filling is the most dynamic stage in rice crops, and it contributes to higher cereal crop yield. According to previous studies, the rate of grain filling is determined by the source activity relative to sink size, carbohydrate accumulation ability, and translocation of assimilates from source tissues to spikelets.[53] It requires a vital source with substantial reserves, and a high sink capability facilitates reserve remobilization from the source to the sink.[54] The restriction of carbohydrate supplements and the inhibition of sink activities were responsible for the low grain weight.[55] The sink strength is controlled by the number and potential size of grains per stem.[56] The source–sink relationship influences the photosynthetic efficiency, grain-filling capacity, and leaf senescence which further influence the cereal yield.

HNT increases carbon loss during night-time respiration and it reduces the activity of the source-sink enzyme, which leads to poor grain filling, decreased biomass, and lower grain weight at the farm and regional levels.[44] Increased carbon loss and decreased sink (grain) enzyme activity are cited as essential factors for lower biomass and yield.[29] HNT had a negative effect on grain quality and rice yield resulting in substantial economic losses,[10] which is similar to the findings of subtropical settings in the Philippines.[7]

Grain filling is a starch-building process and it makes up almost 90% of the weight of a rice grain.[57] SuSase is a rate-limiting enzyme in grain filling because it catalyzes the conversion of sucrose into fructose and uridine diphosphate glucose during the first step of starch synthesis.[50] The ability of rice grains to synthesize starch was also discovered to be substantially related to grain filling rate, which determines final grain weight.[58] Several variables, including floury endosperm-4 (FLO4), starch branching enzyme-I (SBE1), and starch phosphorylase-L (PHOl), can regulate starch production. The variation in the sink activity of rice grains may be explained by investigating the activity and expression of the target enzyme during the grain-filling stage.[59] Apart from the aforementioned enzymes various other enzymes were also regulating the metabolic pathway and have been linked to starch accumulation. The starch synthase IIa genes have spontaneous variation effects on the enzyme and starch properties of some rice cultivars.[60] AGP has also been shown to have a substantial role in the formation of rice endosperm starch,[61] and it has been hypothesized that changes in AGP activity will affect how rice seeds' endosperm filling takes on different forms.[62] Enzymes such as cell wall invertases (cwINV) and sucrose synthase (Susy2) play a crucial role in providing carbon to the filial tissues by breaking sucrose down into UDP-glucose and fructose. The small subunit of (AGPS2b) ADP-glucose pyrophosphorylase catalyzes the conversion of glucose-1-phosphate and ATP into ADP-glucose, a direct substrate for starch manufacture. Both branching enzyme (BEIIb) and granule-bound starch synthase (GBSSI) are required to synthesize amylose and amylopectin, respectively.

Rice grain weight loss during heat stress may be caused by the down-regulation of genes involved in sucrose and starch metabolism.[63] Furthermore, AGPase and SSS play essential roles in starch synthesis in rice grain endosperm, as they use ATP and glucose-1-phosphate (G-1-P) to produce the substrates ADPglucose, the sugar nucleotide used by starch synthase.[64] However, the main function of SSS and SBE is to catalyze the synthesis of amylopectin.[65] According to findings, the endosperm of superior spikelets of rice panicles contains more amylose starch than inferior spikelets.[66] Additionally, SS activity was evaluated as a possible indication for high starch filling, and grain yield in rice[54], and low enzyme activity has been found an impact on the sink strength of rice endosperm.[56] Furthermore, Ishimaru et al.[67] suggested that the weak or delayed activation of the AGP and SS genes may be the cause of the inferior spikelets of rice panicles and insufficient sink strength. The temporal expression pattern of all the key enzymes was involved in rice endosperm and starch synthesis pathways during the grain-filling phase.[58] The rice grain filling is significantly influenced by the restriction of grain sink strength brought on by insufficient starch synthesis. Lesser rice panicle spikelets fill poorly and accumulate the extra assimilates distributed to their advantage because they are unable to synthesize enough starch.[68] Enzyme activity is positively associated with the grain filling rates and endosperm cell division and its activity decreased as seed size dropped due to genetic limitations or an unfavorable placement on the panicle axis. Furthermore, NT and elevated e (CO_2) levels are expected to rise concurrently in present and future climates, implying that rice would eventually be exposed to (CO_2) and HNT conditions. It is believed that the best way to reduce the detrimental effects of HNT on quality and grain yield is to improve crop capacity to satisfy the elevated carbon demand brought on by enhanced night respiration.[51]

6.6 EFFECT ON METABOLIC CHANGES IN THE SOURCE (LEAF), SINK (SPIKE), AND TRANSPORT (STEM) TISSUE

The physiological basis for the production of dry matter in crop plants is governed by the source-sink concept, where the source is the potential for photosynthesis and the sink is the potential for utilizing the products of photosynthesis. A small sink prevents a high yield, whereas a large sink prevents an increase in yield if the source capacity is constrained. Since there is a big variation between tropical and subtropical climates in the number of spikelets and grains, the source looks to be the main obstacle to higher rice yields. The sink poses a constraint, particularly for japonicas, as practically all spikelets in temperate temperatures develop into grains. The source–sink relationship is crucial for crop plant growth and performance.[69] The source strength of a plant is mostly determined by its leaves' morphological and anatomical characteristics as well as photosynthesis per unit leaf area.[70] Similarly, the sink strength is primarily determined by its ability to utilize photosynthates efficiently. The number of spikelets per panicle, tillers, and seeds, as well as sink activity was regulated by the various metabolic enzymes such as cell wall invertase (cwINV) and sucrose synthase (SUS).[71] Functions of the key carbohydrate metabolic enzymes, such as ADP-glucose pyrophosphorylase (AGPase), sucrose phosphate synthase (SPS), sucrose phosphate phosphatase (SPP), and sucrose synthase (SUS) activities are known to be important determinants of the metabolic status of source leaves.[72]

Although several investigations have been done on the detrimental effect of terminal heat stress on wheat yields.[73] Studies on the effects of heat stress on wheat cultivars with various levels of thermotolerance during the grain-filling stage discovered that the metabolism of glucose and TCA-related metabolites decreased in response to heat stress, whereas the amounts of free amino acids were increased.[74] Bread wheat underwent postanthesis heat stress, and levels of arginine, tryptophan, pipecolate, histidine, and α-aminoadipate significantly increased (sources) however, compared to control plants, the levels of dimethyl maleate, anthranilate glycerone, guanine, and galactosyglycerol reduced in heat-stressed plants. Metabolites that increased in heat-stressed plants, such as pipecolate and arginine, had strong negative effects on yield, chlorophyll content and PSII efficiency. In contrast, metabolites that decreased in heat-stressed plants, such as galactosyglycerol and dimethyl maleate, had positive effects on yield and PSII efficiency.[75] In winter wheat, during HNT, there was a substantial increase in proteinogenic amino acids, such as glycine and methionine, as well as particular carbohydrates, such as sucrose, glucose, fructose, myoinositol, raffinose, and maltose in the spikes (sinks). This increase was done to better understand the effects of postheading HNT on sink-source metabolic changes.[29] In contrast to the heat-sensitive TX86A5606 under HNT, the heat-tolerant *Tascosa* displayed a considerable decrease in TCA intermediate substances, such as isocitrate and fumarate, in addition to amino acids, such as lysine, leucine, and tyrosine, in the leaves (sources). These alterations in the buildup of amino acids and the intermediates of the TCA cycle suggested that stress-induced protein and starch breakdown was sped up to accommodate the increased respiratory demand under HNT. During HNT, the altered carbohydrate metabolism benefited the heat-tolerant *Tascosa* in maintaining cell turgor, protecting the electron transport chain, and stabilizing cell membranes.[29]

6.7 HNT-INDUCED CHANGES IN CARBOHYDRATE METABOLISM

Carbohydrates are the end product of photosynthetic CO_2 assimilation. In contrast to photosynthesis, however, the constraint of carbohydrate metabolism is not regulated by a single factor but rather by converting readily assimilable photon donors into sucrose and starch moieties. To protect plant tissue from irreversible harm and to promote energy metabolism, growth, development, photosynthesis, and carbohydrate metabolism must be tightly regulated in a temperature-changing environment.[76]

Numerous studies have been done on the detrimental effects of environmental changes such as stress from high and low temperatures. There is evidence that the metabolism of plant carbohydrates is adversely affected by HT.[77] Due to their persistence over prolonged time in crop cycles, rising NT could potentially have a greater global impact than rising DTs. These temperatures could influence a variety of growth and developmental processes.[48] HNT promotes yield losses through a variety of physiological mechanisms, such as higher night-time respiration, a decreased carbohydrate pool, and lower biomass in field-grown rice, as conflicting with increased spikelet sterility under high DT stress.[48] HNT altered the carbon balance of the entire plant, promoted night-time respiration, altered the metabolism of TCA cycle-related metabolites in the leaves, and altered the carbohydrate metabolisms in wheat spikes.[29] Field-grown rice has shown HNTs impact on the enzymatic activities related to the metabolism of grain starch.[7] The thermodynamics of membrane transport and enzyme activity is directly impacted by a change in temperature regime. Increased enzyme reaction rates would occur as temperatures rose because of thermodynamic constraints if there were no regulations. For instance, this can throw off the equilibrium between the primary and secondary photosynthetic processes, which might lead to the production of oxidative stress markers.

To stabilize photosynthetic CO_2 uptake, rapid heat sensing, signaling, and significant temperature variations were generally essential.[78] A high temperature increases the plasma membrane fluidity, which activates Ca^{2+} channels and increases Ca^{2+} influx into the cell.[79] Then, calcium signaling, kinase phosphatase activation, ROS signaling, transcription factor activation, and phytohormonal regulatory cascades translate the intracellular signals.[80] It triggers the regulatory network that is partially conserved across many species' groupings.[81] In the heat stress response, heat stress factors (HSFs) are crucial regulatory proteins, that collaborate with heat shock proteins (HSPs) to create and activate signaling cascades that stabilize proteins and prevent an excessive accumulation of unfolded or damaged proteins.[82] Thus, clock-dependent photosynthesis connects both temperature sensing and signaling directly to the physiological output.[83]

High temperatures may also have different effects on rates of CO_2 assimilation and photosynthetic electron movement depending on the intensity and length of heat treatment. Comparing the C_3, C_4, and CAM pathways may help researchers understand the thermotolerance of photosynthetic CO_2 absorption and glucose metabolism. The rate of photorespiration increases with temperature, whereas increased CO_2 content in the atmosphere has the opposite effect.[84] Decreased photosynthesis altered carbohydrate contents in stressed plant leaves, whereas changes in sucrose-metabolizing enzyme activity altered carbohydrate metabolism and the contents of fructose, glucose, and sucrose in stressed plants. Plants subjected to HT may have a greater need for energy due to carbon drop at night and fluctuating expression of enzymes catalyzed in metabolic pathways and photosynthesis.[85] However, in contrast, other investigations discovered a drop in the total quantity of glucose, fructose, and sucrose by mild heat.[86] Less ambiguously, starch levels were discovered to be reduced by heat, which may indicate a conserved redirection of photosynthates from accumulation to stress-defensive substances.[87] The final factor contributing to greater electrolyte leakage in heat-treated plants is prolonged heat exposure, which decreases membrane lipid and enhances lipid breakdown.[88] Because environmental stress changes the flow of carbohydrates from source organs to sink organs, soluble sugars are also particularly susceptible to environmental stress.[89] According to previous studies, soluble sugars operate as an Osmo protectant by controlling osmotic adjustment, protecting membranes, and scavenging damaging oxidative stress markers under various types of stress.[90] A recent study of rice seedlings showed the opposite of the assumption made in earlier studies that drought stress increases the quantity of soluble sugar.[91]

6.8 HNT-INDUCED ACTIVATION OF TCA CYCLE AND AMINO ACID METABOLISM IN LEAF

One of the most critical respiratory physiological processes is the TCA cycle. The TCA cycle is a series of catabolic processes which facilitate ATP biosynthesis. The mitochondria contain eight enzymes that carry out the TCA cycle, which combines the production of NADH for the respiratory chain oxidation due to pyruvate and malate oxidation to CO_2.[92] Metabolite profiling is a crucial approach for identifying HNT-related changes in biosynthetic processes, which also identifies compounds that confer endurance and adaptation response and metabolites of stress physiology and cell signaling.[25]

It has been reported that the TCA pathway and amino acid production were activated when rice plants were exposed to HNT, especially in susceptible.[93] Glaubitz et al.[93] reported that the TCA cycle intermediates such as malic acid and other organic acids were increased in the HNT-sensitive rice cultivars. They also observed under HNT, metabolite in medium and resistant cultivars exhibited lower or no changes. In contrast, concentrations of nine amino acids were substantially elevated in susceptible rice varieties and comparable increases in intermediate malic acid within the TCA cycle. Enhanced protein breakdown may cause increasing amino acid concentrations following stress treatment. Consequently, the total protein content was tested; nevertheless, no discernible differences between HNT and control conditions were revealed.

6.9 CONCLUSIONS

The intensity of the NT affected various physiological, biochemical, and molecular indices. Plants' peak physiology and growth occurred primarily at NT higher than the ambient temperature, particularly for leaf growth. Several studies suggest that diurnal variations in vegetation responses to night temperature changes are essential for understanding future changes in vegetation photosynthetic activity and growth. Numerous crops demonstrated significant variation in HNT-induced increases in flag leaf night respiration and acclimation after prolonged HNT exposure. A more substantial decrease in biomass, yield, and related traits under HNT compared to changes in flag leaf carbon balance emphasizes the significance of considering carbon balance at the whole plant level. Fluctuations in carbohydrate metabolism in spikes and the TCA cycle in leaves reveal new information about metabolites involved in HNT response. Therefore, future research on the impact of HNT on crop yield and quality will need to consider night respiration, carbon balance, starch-protein balance, and enzymatic responses.

6.10 WAY FORWARD

It is imperative to conduct new field studies with various increased temperatures throughout the day, night, and multiple seasons on different plant functional groups. Such studies will provide novel insights into the eco-physiological implications of seasonal variations in NT and can be used to enhance the functionality of existing land surface models. For accurate projections of changes in food production, carbon sequestration, and climate regulation, it is important to unravel plant's functional type-specific response patterns to variations in NT. It is an attractive area for future study to examine the genetic regulation of source-sink and functional stay-green carbohydrate interactions at night. Using modern breeding and molecular methods to study the genetic control of essential sink enzymes such as CWI, VI, and SuSy is a step toward creating superior cultivars with HNT tolerance. Additionally, crop models can accurately estimate HNT-induced damage under potential future warming scenarios due to the mechanistic responses discovered under realistic field conditions. Future research should focus on various themes, including how NT affects the allocation and partitioning of photosynthates from source to sink, the expression of genes involved in secondary metabolite biosynthesis and degradation, and the uptake and transport of mineral elements.

KEYWORDS

- **high night temperature**
- **climate change**
- **photosynthesis**
- **respiration**
- **carbon balance**
- **yield**
- **photoassimilate**

REFERENCES

1. IPCC. AR5 Climate Change. Impacts, Adaptation, and Vulnerability; Cambridge Univ. Press: Cambridge, UK, **2014**.
2. FAO. FAOSTAT Database 2014. http://faostat.fao.org/ (accessed on 2 Nov 2019).
3. Sillman, J.; Kharin, V. V.; Zhang, X.; Zwiers, F. W.; Bronaugh, D. Climate Extremes Indices in the CMIP5 Multimodel Ensemble. *J. Geophys. Res. Atmos.* **2013,** *118*, 1716–1733.

4. Couee, I.; Sulmon, C.; Gouesbet, G. E. L.; Amrani, A. Involvement of Soluble Sugars in Reactive Oxygen Species Balance and Response to Oxidative Stress in Plants. *J. Exp. Bot.* **2006,** 449–459.
5. Coast, O.; Ellis, R. H.; Murdoch, A. J.; Quinones, C.; Jagadish, K. S. V. High Night Temperature Induces Contrasting Responses for Spikelet Fertility, Spikelet Tissue Temperature, Flowering Characteristics and Grain Quality in Rice. *Funct. Plant Biol.* **2015,** *42*, 149–161.
6. Peng, S.; Huang, J.; Sheehy, J. E.; Laza, R. C.; Visperas, R. M.; Zhong, X. et al. Rice Yields Decline with Higher Night Temperature from Global Warming. *Proc. Natl. Acad. Sci.* **2004,** *101*, 9971–9975.
7. Bahuguna, R. N. et al. Post-Flowering Night Respiration and Altered Sink Activity Account for High Night Temperature-Induced Grain Yield and Quality Loss in Rice (*Oryza sativa* L.). *Physiol. Plant.* **2017,** *159*, 59–73.
8. Shi, W.; Yin, X.; Struik, P. C.; Solis, C.; Xie, F.; Schmidt, R. C.; Huang, M.; Zou, Y.; Ye, C.; Jagadish, S. V. K. High Day- and Night-Time Temperatures Affect Grain Growth Dynamics in Contrasting Rice Genotypes. *J. Exp. Bot.* **2017,** *68,* 5233–5245.
9. Welch, J. R.; Vincent, J. R.; Auffhammer, M.; Moya, P. F.; Dobermann, A.; Dawe, D. Rice Yields in Tropical/Subtropical Asia Exhibit Large But Opposing Sensitivities to Minimum and Maximum Temperatures. *Proc. Natl. Acad. Sci. USA.* **2010,** *107*, 14562–14567.
10. Lyman, N. B.; Jagadish, S. V. K.; Nalley, L. L.; Dixon, B. L.; Siebenmorgen, T. Neglecting Rice Milling Yield and Quality Underestimates Economic Losses from High-Temperature Stress. *PLoS One* **2013,** *8*, e72157.
11. Zheng, S. H.; Nakamoto, H.; Yoshikawa, K.; Furuya, T.; Fukuyama, M. Influences of High Night Temperature on Flowering and Pod Setting in Soybean. *Plant Prod. Sci.* **2002,** *5*, 215–218.
12. Paembonan, S. A.; Hagihara, A.; Hozumi, K. Long-Term Respiration in Relation to Growth and Maintenance Processes of the Aboveground Parts of a Hinoki Forest Tress. *Tree Physiol.* **1992,** 101–110.
13. McDonald, A. E.; Vanlerberghe G. C. Alternative Oxidase and Plastoquinol Terminal Oxidase in Marine Prokaryotes of the Sargasso Sea. *Gene* **2005,** *349*, 15–24.
14. Kreiner, M.; Harvey, L. M.; McNeil. Oxidative Stress Response of a Recombinant *Aspergillus niger* to Exogenous Menadione and H_2O_2 Addition. *Enzyme Microb. Technol.* **2002,** *30*, 346–353.
15. Christensen, J. H.; Hewitson, B.; Busuioc, A.; Chen, A.; Gao, X.; Held, R.; et al. Regional Climate Projections. In *Climate Change: The Physical Science Basis. Contribution of Working Group I to the Fourth Assessment Report of the Intergovernmental Panel on Climate Change*; Solomon, S. D., Qin, M., Manning, Z., Chen, M., Marquis, K. B., Averyt, M. et al.; Cambridge University Press: Cambridge, 2007; pp 847–940.
16. Gibson, L. R.; Mullen, R. E. Soybean Seed Composition Under High Day and Night Growth Temperatures. *J. Am Oil Chem Soc.* **1996,** *73*, 733–737.
17. Mohammed, A. R.; Tarpley, L. Impact of High Night-time Temperature on Respiration, Membrane Stability, Antioxidant Capacity and Yield of Rice Plants. *Crop Sci.* **2009,** *49*, 313–322.

18. Liang, J.; Xia, J.; Liu, L.; Wan, S. Global Patterns of the Responses of Leaf-Level Photosynthesis and Respiration in Terrestrial Plants to Experimental Warming. *J. Plant Ecol.* **2013,** *6* (6), 437–447.
19. Swain, P.; Baig, M. J.; Nayak, S. Maintenance Respiration of *Oryza sativa* Leaves at Different Growth Stages as Influenced by Nitrogen Supply. *Biol. Plantar.* **2000,** *43*, 587–590.
20. Singh, A.; Kumar, P.; Sharma, M.; Tuli, R.; Dhaliwal, H. S.; Chaudhury, A.; Pal, D.; Roy, J. Expression Patterns of Genes Involved in Starch Biosynthesis During Seed Development in Bread Wheat (*Triticum aestivum*). *Mol. Breed.* **2015,** *35*, 1–9.
21. Morita, S.; Yonemaru, J. I.; Takanashi, J. I. Grain Growth and Endosperm Cell Size Under High Night Temperatures in rice (*Oryza sativa* L.). *Ann. Bot.* **2005,** *95*, 695–701.
22. Cheng, W.; Sakai, H.; Yagi, K.; Hasegawa, T. Combined Effects of Elevated [CO_2] and High Night Temperature on Carbon Assimilation, Nitrogen Absorption, and the Allocations of C and N by Rice (*Oryza sativa* L.). *Agric. For. Meteor.* **2010** *150*, 1174–1181.
23. Djanaguiraman, M.; Prasad, P. V.; Schapaugh, W. T. High Day-or Night-time Temperature Alters Leaf Assimilation, Reproductive Success, and Phosphatidic Acid of Pollen Grain in Soybean [(*Glycine max* L.) Merr]. *Crop Sci.* **2013,** *53*, 1594–1604.
24. Pinto, R. S.; Molero, G.; Reynolds, M. P. Identification of Heat Tolerant Wheat Lines Showing Genetic Variation in Leaf Respiration and Other Physiological Traits. *Euphyt* **2017,** *213*, 1–15.
25. Shulaev, V.; Cortes, D.; Miller, G.; Mittler, R. Metabolomics for Plant Stress Response. *Physiol. Plant.* **2008,** *132*, 199–208.
26. Li, X.; Lawas, L. M.; Malo, R.; Glaubitz, U.; Erban, A.; Mauleon, R.; Heuer, S.; Zuther, E.; Kopka, J.; Hincha, D. K.; Jagadish, K. S.; Metabolic and Transcriptomic Signatures of Rice Floral Organs Reveal Sugar Starvation as a Factor in Reproductive Failure Under Heat and Drought Stress. *Plant Cell Environ.* **2015,** *38*, 2171–2192.
27. IPCC. Technical Summary. In: *Climate change : The Physical Science Basis. Contribution of Working Group I to the Fourth Assessment Report of the Intergovernmental Panel on Climate Change*; Solomon, S., Qin, D., Manning, M., Chen, Z., Marquis, M., Averyt, K. B., Tignor, M., Miller, H. L; Cambridge University Press: Cambridge, 2007; pp 847–940.
28. Singh, R. P.; Prasad, P. V. V.; Sunita, K.; Giri, S. N.; Reddy, K. R. Influence of High Temperature and Breeding for Heat Tolerance in Cotton: A Review. *Adv. Agron.* **2007,** *93*, 313–386.
29. Impa, S. M. et al. Carbon Balance and Source–Sink Metabolic Changes in Winter Wheat Exposed to High Night-Time Temperature. *Plant Cell Environ.* **2019,** *42*, 1233–1246.
30. Rashid, F. A. A.; Scafaro, A. P.; Asao, S.; Fenske, R.; Dewar, R. C.; Masle, J.; Taylor, N. L.; Atkin, O. K. Diel-and Temperature-Driven Variation of Leaf Dark Respiration Rates and Metabolite Levels in Rice. *New Phytol.* **2020,** *228*, 56–69.
31. Frantz, J. M.; Cometti, N. N.; Bugbee, B. Night Temperature Has a Minimal Effect on Respiration and Growth in Rapidly Growing Plants. *Ann. Bot.* **2004,** *94*, 155–166.

32. Mohammed, A. R.; Tarpley, L. Effects of High Night Temperature and Spikelet Position on Yield-Related Parameters of Rice (*Oryza sativa* L.) Plants. *Eur. J. Agron.* **2010,** *33* (2), 117–123.
33. Warrag, M. O.; Hall, A. E. Reproductive Responses of Cowpea (*Vigna unguiculata* (L.) Walp.) to Heat Stress: II. Responses to Night Air Temperatures. *Field Crops Res.* **1984,** *8*, 17–33.
34. Ahmed, F. E.; Hall, A. E. Heat Injury During Early Floral Bud Development in Cowpea. *Crop Sci.* **1993** *33* (4), 764–767.
35. Prasad, P. V. V.; Pisipati, S. R.; Ristic, Z.; Bukovnik, U.; Fritz, A. K. Impact of Night-Time Temperature on Physiology and Growth of Spring Wheat. *Crop Sci.* **2008,** *48*, 2372–2380.
36. Prasad, P. V.; Djanaguiraman, M. High Night Temperature Decreases Leaf Photosynthesis and Pollen Function in Grain Sorghum. *Funct. Plant Biol.* **2011,** *38* (12), 993–1003.
37. Dong, C.; Yuming, F.; Guanghui, L.; Hong, L. Low Light Intensity Effects on the Growth, Photosynthetic Characteristics, Antioxidant Capacity, Yield and Quality of Wheat (*Triticum aestivum* L.) at Different Growth Stages in BLSS. *Adv. Space Res.* **2014,** *53*, 1557–1566.
38. Warrag, M. O. A.; Hall, A. E. Reproductive Responses of Cowpea (*Vigna unguiculata* (L.) Walp.) to Heat Stress: II—Responses to Night Air Temperature. *Field Crops Res.* **1984,** *8*, 17–33.
39. Peet, M. M.; Bartholomew, M.; Effect of Night Temperature on Pollen Characteristics, Growth and Fruit Set in Tomato. *J. Am. Soc. Hort. Sci.* **1996,** *121*, 514–519.
40. Yoon, S. T.; Hoogenboom, G.; Flitcroft, I.; Bannayan, M. Growth and Development of Cotton (*Gossypium hirsutum* L.) in Response to CO_2 Enrichment Under Two Different Temperature Regimes. *Environ. Exp. Bot.* **2009,** *67*, 178–187.
41. Soliz, L. M. A.; Oosterhuis, D. M.; Coker, D. L.; Brown, R. S. Physiological Response of Cotton to High Night Temperature. *Am. J. Plant Sci. Biotechnol.* **2008,** *2*, 63–68.
42. García, G. A.; Dreccer, M. F.; Miralles, D. J.; Serrago, R. A. High Night Temperatures During Grain Number Determination Reduce Wheat and Barley Grain Yield: A Field Study. *Glob. Chang. Biol.* **2015,** *21*, 4153–4164.
43. Song, L.; Jiang, Y.; Zhao, H.; Hou, M. Acquired Thermotolerance in Plants. *Plant Cell Tissue Organ Cult.* **2012,** *111*, 265–276.
44. Shi, W.; Muthurajan, R.; Rahman, H.; Selvam, J.; Peng, S.; Zou, Y.; Jagadish, K. S. Source-Sink Dynamics and Proteomic Reprogramming Under Elevated Night Temperature and Their Impact on Rice Yield and Grain Quality. *New Phytol.* **2013,** *197*, 825–837.
45. Jing, P.; Wang, D.; Zhu, C.; Chen, J. Plant Physiological, Morphological and Yield-Related Responses to Night Temperature Changes Across Different Species and Plant Functional Types. *Front. Plant Sci.* **2016,** *7*, 1774.
46. Echer, F. R.; Oosterhuis, D. M.: Loka, D. A.; Rosolem, C. A. High Night Temperatures During the Floral Bud Stage Increase the Abscission of Reproductive Structures in Cotton. *J Agron Crop Sci.* **2014,** *200*, 191–198.
47. Mohammed, A. R.; Tarpley, L. Effects of High Night Temperature and Spikelet Position on Yield-Related Parameters of Rice (*Oryza sativa* L.) Plants. *Eur. J. Agron.* *33* (2), 117–123.

48. Jagadish, S. V. K.; Murty, M. V. R.; Quick, W. P. Rice Responses to Raising Temperatures-Challenges, Perspectives and Future Directions. *Plant Cell Environ.* **2015,** *38*, 1686–1698.
49. Frottin, F.; Espagne, C.; Traverso, J. A.; Mauve, C.; Valot, B.; Lelarge-Trouverie, C.; Zivy, M.; Noctor, G.; Meinnel, T.; Giglione, C. Cotranslational Proteolysis Dominates Glutathione Homeostasis to Support Proper Growth and Development. *Plant Cell* **2009,** *21* (10), 3296–3314.
50. Fu, J.; Huang, Z. H.; Wang, Z. Q.; Yang, J. C.; Zhang, J. H.; Pre-Anthesis Nonstructural Carbohydrate Reserve in the Stem Enhances the Sink Strength of Inferior Spikelets During Grain Filling of Rice. *Field Crop Res.* **2011,** *123*, 170–182.
51. Impa, S. M.; Vennapusa, A. R.; Bheemanahalli, R.; Sabela, D.; Boyle, D.; Walia, H.; Jagadish, S. V. K. High Night Temperature Induced Changes in Grain Starch Metabolism Alters Starch, Protein, and Lipid Accumulation in Winter Wheat. *Plant Cell Environ.* **2020,** *43*, 431–447.
52. Wang, W.; Cui, W., X.; Gao, H.; Wei, H.; Zhang, H. Effects of Early- and Late-Sowing on Starch Accumulation and Associated Enzyme Activities During Grain Filling Stage in Rice. *Rice Sci.* **2021,** *28*, 191–199.
53. Zhao, B. H.; Wang, P.; Zhang, H.; Zhu, Q.; Yang, J. Sourcesink and Grain Filling Characteristics of Two-Line Hybrid Rice Yangliangyou 6. *Rice Sci.* **2006,** *13*, 34–42.
54. Counce P. A.; Gravois, K. A. Sucrose Synthase Activity as a Potential Indicator of High Rice Grain Yield. *Crop Sci.* **2006,** *46*, 1501–1507.
55. Yang, J.; Zhang, J. Grain-Filling Problem in "Super" Rice. *J. Exp. Bot.* **2010,** *61*, 1–5.
56. Patel, R.; Mohapatra, P. K. Assimilate Partitioning Within Floret Components of Contrasting Rice Spikelets Producing Qualitatively Different Types of Grains. *Austr. J. Plant Physiol.* **1996,** *23*, 85–92.
57. Duan, M.; Sun, S. Profiling the Expression of Genes Controlling Rice Grain Quality. *Plant Mol. Biol.* **2005,** *59*, 165.
58. Liang, J.; Zhang, J.; Cao, X. Grain Sink Strength May Be Related to the Poor Grain Filling of Indica-Japonica Rice (*Oryza sativa*) Hybrids. *Physiol. Plant.* **2001,** *112*, 470–477.
59. You, C.; Zhu, H.; Xu, B.; Huang, W.; Wang, S.; Ding, Y.; Liu, Z.; Li, G.; Chen, L.; Ding, C.; Tang, S. Effect of Removing Superior Spikelets on Grain Filling of Inferior Spikelets in Rice. *Front. Plant Sci.* **2016,** *7*, 1161.
60. Umemoto, T.; Aoki, N.; Lin, H.; Nakamura, Y.; Inouchi, N.; Sato, Y.; Yano, M.; Hirabayashi, H.; Maruyama, S. Natural Variation in Rice Starch Synthase IIa Affects Enzyme and Starch Properties. *Funct. Plant Biol.* **2004,** *31* (7), 671–684.
61. Nakamura, Y.; Yuki, K. Changes in Enzyme Activities Associated with Carbohydrate Metabolism During the Development of Rice Endosperm. *Plant Sci.* **1992,** *82* (1), 15–20.
62. Kato, T.; Shinmura, D.; Taniguchi, A. Activities of Enzymes for Sucrose-Starch Conversion in Developing Endosperm of Rice and Their Association with Grain Filling in Extra-Heavy Panicle Types. *Plant Prodn. Sci.* **2007,** *10*(4):442–450.
63. Yamakawa, H.; Hirose, T.; Kuroda, M.; Yamaguchi, T. Comprehensive Expression Profiling of Rice Grain Filling Related Genes Under High Temperature Using DNA Microarray. *Plant Physiol.* **2007,** *144*, 258–277.

64. Stitt, M.; Zeeman, S. C. Starch Turnover: Pathways, Regulation and Role in Growth. *Curr. opin. Plant Biol.* **2012,** *15* (3), 282–292.
65. Jiang, H.; Dian, W.; Wu, P. Effect of High Temperature on Fine Structure of Amylopectin in Rice Endosperm by Reducing the Activity of the Starch Branching Enzyme. *Phytochem.* **2003,** *63* (1), 53–59.
66. Matsue, Y.; Odahara, K.; Hiramatsu, M. Differences in Amylose Content, Amylographic Characteristics and Storage Proteins of Grains on Primary and Secondary Rachis Branches in Rice. *Jpn. J. Crop Sci.* **1995,** *64* (3), 601–606.
67. Ishimaru, T.; Hirose, T.; Matsuda, T.; Goto, A.; Takahashi, K.; Sasaki, H.; Terao, T.; Ishii, R. I.; Ohsugi, R.; Yamagishi, T. Expression Patterns of Genes Encoding Carbohydrate-Metabolizing Enzymes and Their Relationship to Grain Filling in Rice (*Oryza sativa* L.): Comparison of Caryopses Located at Different Positions in a Panicle. *Plant Cell Physiol.* **2005,** *46* (4), 620–628.
68. Moncada, P.; Martinez, C. P.; Borrero, J.; Châtel, M.; Gauch Jr, H.; Guimaraes, E.; Tohmé, J.; McCouch, S. R. Quantitative Trait Loci for Yield and Yield Components in an *Oryza sativa*× *Oryza rufipogon* BC2F2 Population Evaluated in an Upland Environment. *Theoret. Appl. Gen.* **2001,** *102* (1), 41–52.
69. White, A. C.; Rogers, A.; Rees, M.; Osborne, C. P. How Can We Make Plants Grow Faster? A Source–Sink Perspective on Growth Rate. *J. Exp. Bot.* **2016,** *67* (1), 31–45.
70. Mathan, J.; Bhattacharya, J.; Ranjan, A. Enhancing Crop Yield by Optimizing Plant Developmental Features. *Development* **2016,** *143* (18), 3283–3294.
71. Smith, M. R.; Rao, I. M.; Merchant, A. Source–Sink Relationships in Crop Plants and Their Influence on Yield Development and Nutritional Quality. *Front. Plant Sci.* **2018,** *9*, 1889.
72. Osorio, S.; Ruan, Y. L.; Fernie, A. R. An Update on Source-to-Sink Carbon Partitioning in Tomato. *Front. Plant Sci.* **2014,** *5*, 516.
73. Zhang, X.; Högy,P.; Wu, X.; Schmid, I.; Wang, X.; Schulze, W. X.; Jiang, D.; Fangmeier, A. Physiological and Proteomic Evidence for the Interactive Effects of Post-Anthesis Heat Stress and Elevated CO_2 on Wheat. *Proteomic* **2018,** *18* (23), 1800262.
74. Wang, X.; Lijiang, H.; Yunze, L.; Bingjin, W.; Xue, G.; Manshuang L.; Jun, W.; Qixin, S.; Elizabeth, V.; Shengbao, X. Metabolic Adaptation of Wheat Grain Contributes to a Stable Filling Rate Under Heat Stress. *J. Exp. Bot.* **2018,** *69*, 5531–5545.
75. Thomason, K.; Babar, M. A.; Erickson, J. E.; Mulvaney, M.; Beecher, C.; MacDonald, G. Comparative Physiological and Metabolomics Analysis of Wheat (*Triticum aestivum* L.) Following Post-Anthesis Heat Stress. *PLoS One* **2018,** *13* (6), e0197919.
76. Rees, T. Compartmentation of Plant Metabolism. In: *The Biochemistry of Plants; Physiology of Metabolism*; Davies, D. D., Ed., Vol. 12; Academic Press: San Diego, 1987; pp. 87–115.
77. Narayanan, S., Prasad, P. V. V. ; Fritz, A. K.; Boyle, D. L.; Gill. B. S.; Impact of High Night-Time and High Daytime Temperature Stress on Winter Wheat. *J. Agron. Crop Sci.* **2015,** *201*, 206–218.
78. Ranty, B.; Aldon, D.; Cotelle, V.; Galaud, J. P.; Thuleau, P.; Mazars, C. Calcium Sensors as Key Hubs in Plant Responses to Biotic and Abiotic Stresses. *Front. Plant Sci.* **2016,** *7*, 327.

79. Sajid, M.; Rashid, B.; Ali, Q.; Husnain. T. Mechanisms of Heat Sensing and Responses in Plants: It Is Not All About Ca^{2+} Ions. *Biol. Plant.* **2018,** *62*, 409–420.
80. Mittler, R.; Finka, A.; Goloubinoff, P. How Do Plants Feel the Heat? *Trend. Biochem. Sci.* **2012,** *37* (3), 118–125.
81. Richter, K.; Haslbeck, M.; Buchner, J. The Heat Shock Response: Life on the Verge of Death. *Mol. Cell* **2010,** *40* (2), 253–266.
82. Qu, A. L.; Ding, Y. F.; Jiang, Q.; Zhu, C. Molecular Mechanisms of the Plant Heat Stress Response. *Biochem. Biophys. Res. Commun.* **2013,** *432*, 203–207.
83. Dodd, A. N.; Kusakina, J.; Hall, A.; Gould, P. D.; Hanaoka, M. The Circadian Regulation of Photosynthesis. *Photosynth. Res.* **2014,** *119*, 181–190.
84. Long, S. P. Modification of the Response of Photosynthetic Productivity to Rising Temperature by Atmospheric CO_2 Concentrations: Has Its Importance Been Underestimated? *Plant Cell Environ.* **1991,** *14*, 729–739.
85. Sharmin, S. A.; Alam, I.; Rahman, M.; Kim, K. H.; Kim, Y. G.; Lee, B. H.; Mapping the Leaf Proteome of *Miscanthus sinensis* and Its Application to the Identification of Heat-Responsive Proteins. *Planta* **2013,** *238*, 459–474.
86. Vasseur, F.; Pantin, F.; Vile, D. Changes in Light Intensity Reveal a Major Role for Carbon Balance in Arabidopsis Responses to High Temperature. *Plant Cell Environ.* **2011,** *34*, 1563–1576.
87. Atanasov, V.; Fürtauer, L.; Nägele, T. Indications for a Central Role of Hexokinase Activity in Natural Variation of Heat Acclimation in *Arabidopsis thaliana. Plants* **2020,** *9*, 819.
88. Tang, T.; Peile, L.; Guowei, Z.; Weiqi, L. Two Phases of Response to Long-Term Moderate Heat: Variation in Thermotolerance Between *Arabidopsis thaliana* and Its Relative *Arabis paniculata. Phytochemical* **2016,** *122*, 81–90.
89. Rosa, M.; Carolina, P.; Griselda, P.; Roque, I.; Juan, A.; González, M. H.; Fernando E. P. Soluble Sugars: Metabolism, Sensing and Abiotic Stress: A Complex Network in the Life of Plants. *Plant Signal. Behav.* **2009,** *4*, 388–393.
90. Parvaiz, A.; Satyawati, S. Salt Stress and Phyto-Biochemical Responses of Plants—A Review. *Plant Soil Environ.* **2008,** *54*, 89.
91. Xu, W.; Kehui, C.; Aihui, X.; Lixiao, N.; Jianliang, H.; Shaobing, P. Drought Stress Condition Increases Root to Shoot Ratio via Alteration of Carbohydrate Partitioning and Enzymatic Activity in Rice Seedlings. *Acta Physiol. Plant.* **2015,** *37*, 1–11.
92. Fernie, A. R.; Trethewey, R. N.; Krotzky, A. J.; Willmitzer, L. Metabolite Profiling: from Diagnostics to Systems Biology. *Nat. Rev. Mol. Cell Biol.* **2004,** *5*, 763–769.
93. Glaubitz, U.; Erban, A.; Kopka, J.; Hincha, D. K.; Zuther, E. High Night Temperature Strongly Impacts TCA Cycle, Amino Acid and Polyamine Biosynthetic Pathways in Rice in a Sensitivity-Dependent Manner. *J. Exp. Bot.* **2015,** *66*, 6385–6397.

CHAPTER 7

High-Temperature Stress: Effects on Plant Metabolism Specific to Lipidome

SAURABH PANDEY[1], SUNIDHI MISHRA[2], and K. K. DWIVEDI[3]

[1]*Department of Agriculture, Guru Nanak Dev University, Amritsar, Punjab, India*

[2]*Department of Horticulture, College of Agriculture, Korea, Indira Gandhi Agriculture University, Raipur, Chhattisgarh, India*

[3]*ICAR- Indian Grassland and Forage Research Institute, Jhansi, Uttar Pradesh, India*

ABSTRACT

Recent global warming trends are alarming and potentially threaten crop production, requiring more attention. IPCC's recent reports indicated more frequent heat waves with longer warm seasons and increased night temperatures. Therefore, developing High temperature (HT) stress-resilient crops is crucial for global crop productivity. HT stress affects the crops at the physiological, biochemical, and molecular levels to alter their productivity. Plants employ defense mechanisms to overcome HT-induced structural alterations and superfluidity. Lipid remodeling by specific genotypes provides such resistance/tolerance to plants. Lipidomics could identify such different lipid molecules involved in HT stress. High-throughput omics techniques such as tandem mass spectrometry (MS) helped in the large-scale lipid molecules screening under the stress conditions in minimum time. Conventional physiological methods for HT-tolerant genotype screening were limited by the drawbacks of selecting permissive

Impact of High Night Temperature on Plant Biology: Toward Sustainable Plant Adaptation to Climate Change. Rajesh Kumar Singhal, Raju Bheemanahalli, Saurabh Pandey, and Pratibha M. D. (Eds.)

genotypes with fewer traits for tolerance. Therefore, selection based on potential lipid biomarkers governing the HT tolerance/resistance with the involvement of multiple genes in remodeling will be helpful. These biomarkers constitute the lipidotype of the plant, which interlinks their genotype to phenotype. Recent quantitative studies through genome-wide association study (GWAS) [metabolome GWAS (mGWAS)] highlighted this link between metabolites, genotypes, and their stress-tolerant phenotypes. In this chapter, we will discuss the plant lipidome, its constituents, membrane lipid synthesis, and different advances in analytical techniques for plant lipidome study during stress which will help to design future generation climate resilience crops.

7.1 INTRODUCTION

Global climate change scenarios increase the frequent abiotic stresses affecting the planet's food security. Abiotic stresses include high temperature (HT), water, heavy metal, salinity, etc. Among them, HT is the most severe threat due to the constant global temperature increase by increased greenhouse gas emissions. Climate changes brought the rise in worldwide temperature and a temperature rise range of 1.5–2°C is estimated to cause warmer seasons, increased night temperatures with frequent heat waves, and extreme crop losses.[1] Plants defend themselves from extreme temperatures by altering their physiology and metabolism. For example, they produce various compatible solutes to maintain their redox balance and osmotic adjustment to organize proteins in the cellular environment under stress.[2,3] One of the crucial physiological changes is the change in membrane fluidity, mainly governed by the alteration in membrane lipid metabolism responsible for maintaining membrane integrity.[4] It was shown in wheat and soybean heat-tolerant genotypes that saturated lipid content increases to avoid superfluidity of the membrane. Therefore evaluation of comprehensive lipid species through lipidomics under high-temperature stress could be a potential tool for understanding plant stress and the development of stress-resilient crops.[5]

Compositional variation of lipid under stress responses has already been shown in wheat, soybean, and Ethiopian mustard.[6–9] The high-temperature stress also affected membrane-bound organelles such as chloroplast, resulting in their lipid composition change leading to membrane damage and reduced photosynthesis.[6] These alterations in lipids are carried out through specific enzymes that coordinate their oxidation, desaturation, glycosylation, and acylation.[6,10] These enzymes also act as biomarkers for screening the heat-tolerant genotypes for hyperfluidity of membrane lipids. Thylakoid membranes of the chloroplast responsible for light reactions in photosynthesis are highly susceptible to high-temperature damage, resulting in decreased PSII quantum yield.[7] Also, high-temperature stress increases the nonbilayer forming lipids in membranes, denaturation of lipids and proteins leading to disruption of lipid–protein interactions and increases the reactive oxygen species (ROS) causing lipid peroxidation. Therefore, maintaining the plant's lipid integrity and fluidity is vital for its survival under high-temperature stress.[11–13] Genetic studies through mutations revealed a close association of membrane lipid biosynthesis, desaturation, and transport with plant stress responses during HTs.[14–16] Thus understanding the possible molecular mechanisms responsible for membrane lipid status modulations under stress will help us to devise strategies for plant tolerance. In this regard, integrating omics techniques, including metabolomics and transcriptomics, will be essential to uncover the metabolism-related genes under stress. However, these traits are complex and follow the quantitative inheritance, which requires robust genomic tools such as metabolome-assisted genome-wide association study (mGWAS) to pinpoint the lipids desired for high-temperature stress tolerance.

This chapter will explore plant lipid diversity, biosynthesis, and lipid quantification methods. Further, we will explore the integrated omics and genetics tools to understand the lipidotype of the different crops concerning their phenotype and associations under stress. Overall this chapter will add to our knowledge about plant lipidome and its crucial role in designing climate-resilient crops.

7.2 PLANT LIPIDS: DIVERSE MOLECULES WITH STRESS TOLERANCE

Lipids are metabolic energy molecules produced through enzymatic reactions, and different factors regulate their metabolism. These do not directly come from the central dogma of life.[17] However, they play an essential role in cellular functions as a part of biological membranes, cell ultrastructures, signaling molecules, and energy sources.[18–20] Maintaining a steady state of lipids is crucial, although they constantly change in response to changing environments. In plants, lipids are involved in photosynthesis structurally, vesicle trafficking, secretion, signal transduction, and cytoskeleton rearrangements.[21,22] These lipids are structurally diverse, abundant in various combinations, and unique to particular cells and tissues.[23] This diversity is attributed to its chemical structures and compositions, which vary between tissues/cells, organelles, and species and are unevenly distributed across the membranes.[24] Also, lipid modifications such as glycosylation, acylation, desaturation, and oxidation increase its diversity.[7] Chemically lipids are made up of fatty acids with head and tail groups to add to their diversity, along with their saturation determined by the number of double bonds.[23] Broadly plant membrane lipids are classified as glycerolipids, sphingolipids, and sterols.[25] Phospholipids have phosphorous in their head group and constitute a significant part of cell membranes, with galactolipids as a major part and phosphatidic acid (PA) as the simplest phospholipid for the precursor to other lipids.[26] Other examples of head groups are glycerol, serine, inositol, ethanolamine, and choline, known as phosphatidyl glycerol (PG), phosphatidyl serine (PS), phosphatidyl inositol (PI), phosphatidyl ethanolamine (PE), and phosphatidylcholine (PC).[24] Interestingly, nonphosphorous galactolipids and sulfolipidsdominate the thylakoid membrane and are essential for photosystem II working.[27,28] Galactolipids are predominant in chloroplast membranes and help to maintain photosynthesis under phosphate-limiting conditions.[29] Similarly, sphingolipids are grouped into ceramides (Cer), glycosyl inositol phosphor ceramides (GIPCer), glucosyl ceramides (GCer), and long-chain bases (LCB).[30,31] Different plant sterols such as campesterol, β-sitosterol, and stigmasterol are present in free or modified forms that regulate membrane fluidity and permeability.[32–34]

Studying these lipids in their entirety requires understanding their roles, alterations during environmental stress, and impacts on the biological system.[35] Here, lipidomics became essential as it deals with the complete characterization of lipids in a system. It also requires sophisticated high-resolution analytical techniques, such as an advanced mass-spectrometry (MS)-based lipidomics platform.[36] Lipidomics deals with the target in two analytical approaches known as targeted and untargeted lipidomics. The targeted approach will analyze specific lipid classes, whereas the entire lipidome will be screened in the untargeted approach.[23,37]

7.3 MASS SPECTROMETRY-BASED LIPID PROFILING

The cellular diversity of lipids made it challenging to profile them by earlier analytical techniques, which are time-consuming and lack sensitivity and resolution for the set of cellular lipids.[23,38,39] With the advent of omics, several analytical methods have been developed to analyze the lipidome of an organism. These include MS, ultraviolet–visible (UV-Vis) spectroscopy, Fourier-transform infrared spectroscopy (FT-IR), and nuclear magnetic resonance (NMR).[17,40] Further, soft-ionization techniques, for example, electrospray ionisation-mass spectrometry (ESI-MS), helped cellular lipidomics.[41] A typical MS system comprises the ion source, analyzer, and detector. Different ion sources can be attached for mass analyzers to widen the lipid identification.[42] These include atmospheric pressure chemical ionization (APCI), electrospray ionization (ESI), desorption ionization (DESI), matrix-assisted laser desorption ionization (MALDI), and direct analysis in real time (DART), which are commonly used for lipid analysis.[43–45] Different analyzers such as ion trap, orbitrap, time of flight (TOF), triple quadrupole (QqQ), and Fourier transform ion cyclotron resonance (FT-ICR) or Fourier transform (FT) were used in MS-based systems.[35,46] MS-based lipidomics analytical systems use platforms of liquid chromatography coupled MS (LC-ESI-MS) and direct infusion MS (ESI-MS).[47,48] In LC-MS, lipids were separated based on class and then proceeded for ionization with specific elution time for different lipids.[41] In direct infusion (shotgun lipidomics), constant concentrations of lipid solution were fed without any separation before the ionization steps. This method is chosen due to its accuracy for different lipid species and short run time for large-scale lipid analysis.[48,49] MS-based lipidomics utilizes separation systems, including hydrophilic interaction chromatography (HILIC) or reverse-phase chromatography (RPC). Separation of various lipid classes depends on their physiochemical properties, and due to their high sensitivity, specificity, and better separation efficiency, LC-MS has been the preferred method. Plant lipidomics uses ESI-MS/MS, also known as triple quadrupole tandem MS, for analyzing and collecting a large amount of spectral data. Different MS spectra allow the identification of specific lipid molecules with fragments originating from the head group as a criterion for screening.[50–52]

Plant high-temperature stress response comprises a change in membrane lipid composition to cope with the stress-generated hyperfluidity.[38]

High-throughput MS platforms are crucial for screening such complex lipid alteration data and have been employed to measure the change in plant membrane lipid profile.[4,5] Lipidomics explores the possibility of the abundance of one or multiple lipids in a sample and their relative quantification and pattern change for characterizing treatment response with biomarker discovery.[42] Quantifying lipid species requires external or internal standards concerning absolute quantification and depends upon the lipid species and its classes.[41,53] Isotope labeling of standards is preferred for the quantification; however, not possible for all the lipid species. Hence nonnative lipid internal standards were commonly used for quantification in MS analysis.[54] Therefore, plant-specific lipid analysis requires high throughput MS platforms for identifying stress-specific membrane lipids responsible for the stress resilience of specific genotypes and cultivars.

7.4 MEMBRANE LIPIDS AND LIPIDOME CHANGES UNDER HIGH TEMPERATURES

To cope with environmental stresses, plants reinforce their plasma membrane, the interface of interacting with the environment. Plasma membranes are universally made of proteins and lipids to act as permeability barriers and maintain cellular homeostasis.[55,56] Upon stress, these structural components are subject to reprogramming and alterations and exposed to stress-induced damage.[57] Therefore, membrane stability, electrolyte leakage, etc., are considered crucial factors for screening high-temperature tolerant plants.[58] These membranes primarily behave as a fluidic bilayer, with specific patches that could transit to gel or nonbilayer under stress.[59,60] Fatty acid chains' polar heads and nonpolar tails are arranged in the membrane bilayer to create the aqueous outer and inner membrane environment.[61] The lipid tail's cis-double bonds maintain optimum membrane fluidity.[62] For example, in the membrane gel phase, the packing of lipids is more closed due to extended fatty acid chains and their dense and ordered packing in contrast to the fluidic phase. Under normal conditions, gel and hexagonal phases have limited functions and exist. In stress conditions, for example, HTs lead to a transition from gel to fluid and fluid to hexagonal phases.[63] These membrane changes lead to ion leakage and cell death under stress conditions. Also, HT leads to oxidative stress and ROS accumulation, leading to membrane lipids' peroxidation and cell death.[64] Glycerolipids of the membrane oxidized to generate cytotoxic molecules, which harm plant photosynthetic proteins' cleavage and aggregation.[65,66] Moreover, thylakoid membranes were also damaged due to high-temperature stress leading to a reduction in photosynthesis.[67] Electron transfer reactions of the light reactions in the thylakoid membrane are essential for photosynthesis, and glycerolipids are the main component of these bilayers.[68] High-temperature stress studies in different plants such as rice, wheat, sorghum, maize, soybean, and cotton suggest net photosynthesis, chlorophyll pigments, electron transport rate, stomatal conductance, and variable fluorescence/maximum fluorescence (F_v/F_m) ratios were reduced.[69–74] Interestingly, the amount of lipid oxidation, desaturation, acylation, and membrane organelles were responsible for photosynthesis reduction and subsequent yield loss.[7]

Plants have developed different strategies for adapting these biochemical and physiological changes in response to high-temperature stress. Galactolipids are essential to maintain the lipid profile and fluidity of the thylakoid membrane.[38,75] Therefore plant alters their lipid composition, saturation levels, and fluidity, known as lipid remodeling, to counter the high-temperature stress. Plants optimize their membrane properties to minimize the temperature effect on photosynthesis and other cellular processes.[49,76] Lipid remodeling in response to HTs changes the number of specific lipids to transition the phase in membranes for their stability.[77] Lipid profile alterations with lipid composition and fatty acid saturation are the particular markers associated with the tolerance and susceptibility of plants.[13,78] Plants change their glycerolipid biosynthesis to effectively remodel the membrane lipid composition under stress.[79] Lipids in plants' membranes exist with double bonds in cis-configuration to create a bend in fatty acid structure to avoid dense lipid packing.[6] High-temperature stress targets this bend in the membrane to destabilize the membrane structure and make it superfluid. Plant responses against temperature stress are specific, which could be attributed to their lipid characteristics—for example, membrane glycerolipid unsaturation level and polyunsaturated fatty acid content decrease in contrast to monounsaturated and saturated fatty acid levels increase for providing temperature tolerance.[6,7] These changes are governed by alterations in the expression of fatty acid desaturases in membrane lipids.[4,8] Associated Tri-acyl glycerol (TAG) production by acyl transferases contributes to lipid remodeling under stress.[80,81] Plants also employ less energy-consuming mechanisms such as glycerolipid pathway balancing or head group turnover to change their lipid composition under temperature stress for optimum membrane function.[82]

Wheat plants under high-temperature stress sequester more saturated lipid species from ER to chloroplast for maintaining the thylakoid membranes. Therefore plants could also alter the biosynthesis of less saturated lipids to increase their pathway contribution, for example, galactolipid synthesis under HTs.[38] The composition of bilayer-forming and nonbilayer-forming lipid molecules and their proportions determine the membrane stability in the thylakoid membrane for their functions under temperature stress.[83] The level of oxidized lipids also acts as an excellent biomarker to screen plants for high-temperature stress.[6] Specific extra-plastidial lipids are also remodeled; for example, an increase in lysophospholipid under stress might indicate plant response.[5] Increasing sterol glycoside or lipids helps maintain the membrane structure by assisting them in phasing transitions.[84] Plants follow different strategies to remodel their membrane lipid composition to preserve its functioning, fluidity, and structure to combat high-temperature stress.

7.5 PLANT LIPIDOTYPE: LINKING THE GENOTYPE TO PHENOTYPE

High-temperature stress is often associated with loss of photosynthesis, its rate, and yield, attributed to organellar damage and membrane lipid modulations shown in wheat crops.[6,7] These lipid alterations are enzyme-dependent oxidation, acylation, and glycosylation with desaturation to alter the membrane.[85] Plant lipidotype changes under high-temperature stress to maintain membrane integrity. However, this remodeling is genotype dependent and therefore varies in efficiency and extant of remodeling under stress.[86,87] We discussed these remodeling strategies in the previous section, and their contribution to plant tolerance requires extensive studies.

Plant temperature stress tolerance is polygenic with genotypic variability for heat stress tolerance.[88,89] These are controlled by epistatic quantitative trait loci (QTLs) or several minor effects traits due to the effect of HT on different cellular processes at different times.[90] Therefore, breeding for high-temperature tolerance requires identified and validated QTLs/minor genes, which are limited in numbers and poorly understood.[91] Also, genotype-environment interaction significantly affects the trait of interest and makes breeding efforts more challenging. There is a need to screen for large-effect QTLs along with markers that could assist the breeding process.[78] Conventional methods have shortcomings in direct selection for high-yield phenotypes due to low heritability and control of heat tolerance by a network of major and minor QTLs.[92] With the lack of screening methods for phenotyping and genotype and environment interactions, this complexity is considered a significant bottleneck for stress resilience crop breeding efforts. Hence, focusing on intermediate traits which are closely related to the stressed plants' physiological and biochemical status could reduce heat tolerance trait complexity.[93,94] Specially physiological trait-based breeding for heat stress tolerance could be ideal for screening as well as trait transfer strategy.[95] These traits include membrane thermostability, higher photosynthetic rates, photosynthetic efficiency, antioxidant potential, reproductive characteristics, canopy structures, and lower respiration rates.[8]

Metabolomes are the study of metabolites that bridge plant genotypes and phenotypes.[93] Utilizing the biomarkers related to plant metabolome for metabolic traits would be ideal for determining the heat-stress tolerant trait.[96] These metabolic networks are complicated due to interconnections with various biochemical pathways and are affected by environmental clues. At the genome level expression of different genes regarding the stress responses determine the quantitative status of metabolites this makes metabolome study difficult.[97] Predominantly, HTs disrupt the plant membranes hampering photosynthesis and metabolism to affect the yield.[97] Therefore silencing of genes related to fatty acid metabolism responsible for saturation levels of membranes could be a possible strategy for plant tolerance as shown in tobacco FAD gene mutants providing high-temperature tolerance.[14]

Plant lipidomics gaining importance for stress due to their impact on membranous structures which is important to perform various cellular functions including photosynthesis. We have discussed lipid remodeling in response to stress which is dependent upon genotypes and has been studied in rice, wheat, grapevine, and soybean.[6,77,98] Complex phenotypes of heat-tolerant genotypes are directly or indirectly associated with specific metabolome profiles which could bridge the gap between phenotype at the field level and genotype.[99,100] Moreover identifying such functional candidates associated with quantitative variation related to lipid remodeling under stress conditions would be easily utilized for breeding programs. For this, we could follow the sequence alignment based on homology and reverse genetics for metabolite/lipid identification.[93] For example, the identification of benzoxazinoid genes in wheat with the help of maize orthologues.[101] However, with less homology, this strategy became less effective. Coexpression analysis is another strategy for the identification of lipid metabolism-related candidates.[102] This strategy requires prior knowledge of gene functions with validated genes. Additional datasets generated from multiomics such as metabolomics, lipidomics, etc. are required to identify candidate genes with forward and reverse genetics approaches.[103,104] Integrated omics strategy has proven to be essential for determining metabolic quantitative trait loci (mQTLs) hotspots in different plants.[105–109] Here, QTL mapping or genome-wide association studies (GWAS) with datasets from metabolomics are used for generating mQTL and mGWAS signifying the associations between chromosomal locations and metabolite quantities in the diversity panel.[110,111] These linkages will be utilized for candidate identification to determine the metabolic variation that could be used for marker-assisted selection in breeding programs.[93,94] mGWAS utilizes larger populations on natural linkage disequilibrium generated by ancestral recombination events already available, whereas mQTL has fewer parental genotypes with limited diversity and recombination events for metabolites.[112,113] This was explored for generating mGWAS in rice, wheat, maize, barley, foxtail millet, and tomato.[105,106,109,111,114,115] Understanding lipidotype for high-temperature tolerance using a diverse population for GWAS with genotype-by-sequencing followed by assessing membrane thermostability and targeted lipidomics could be a better strategy for identifying lipidotype QTLs (lQTLs). This will assist in developing markers that could be utilized for breeding stress resilience crops.

7.6 FUTURE PROSPECTS

With climate change and global warming, both day and night temperatures increase are constant and affect crop productivity. The complex nature of high-temperature stress tolerance traits requires omics-related large-scale data to identify candidates responsible for lipid remodeling. The plant has diverse lipid molecules responsible for membrane structures performing various cellular functions. Characterizing such lipid variation requires high-throughput methods targeting a specific class of lipids. Heat stress-induced changes in plants' lipid composition make them an exciting candidate for investigation. The plant utilizes lipid desaturation and other modifications to tolerate the high-temperature stress by remodeling its membrane lipids fluidity, structure, and chemical nature. However, this field needs more exploration to understand the possible pathways affected and specific lipid classes and their concerned enzymes responsible for such changes under stress conditions. Integration of omics tools with the plant metabolome could provide valuable data linking the plant genotype to the phenotype and identify underlying candidates and markers that could be utilized for generating future generation climate resilience crops with more accuracy under breeding programs.

KEYWORDS

- **lipidome**
- **high-temperature stress**
- **night temperature stress**
- **metabolites**
- **lipidomics techniques**

REFERENCES

1. IPCC. Summary for Policymakers. In: *Global Warming of 1.5°C: IPCC Special Report on Impacts of Global Warming of 1.5°C above Pre-industrial Levels in Context of Strengthening Response to Climate Change, Sustainable Development, and Efforts to Eradicate Poverty*; IPCC, Ed.; Cambridge University Press: Cambridge, 2022; pp 1–24.
2. Hanif, S.; Saleem, M. F.; Sarwar, M.; Irshad, M.; Shakoor, A.; Wahid, M. A.; Khan, H. Z. Biochemically Triggered Heat and Drought Stress Tolerance in Rice by proline Application. *J. Plant Growth Regul.* **2021,** *40*, 305–312.
3. Bhardwaj, A.; Devi, P.; Chaudhary, S.; Rani, A.; Jha, U. C.; Kumar, S.; Bindumadhava, H.; Prasad, P.; Sharma, K. D.; Siddique, K. H. 'Omics' Approaches in Developing Combined Drought and Heat Tolerance in Food Crops. *Plant Cell Rep.* **2021,** *41*, 699–739.

4. Higashi, Y.; Okazaki, Y.; Myouga, F.; Shinozaki, K.; Saito, K. Landscape of the Lipidome and Transcriptome Under Heat Stress in *Arabidopsis thaliana. Sci. Rep.* **2015,** *5*, 10533.
5. Welti, R.; Li, W.; Li, M.; Sang, Y.; Biesiada, H.; Zhou, H. E.; Rajashekar, C. B.; Williams, T. D.; Wang, X. Profiling Membrane Lipids in Plant Stress Responses: Role of Phospholipase Dα in Freezing-Induced Lipid Changes in *Arabidopsis*. *J. Biol. Chem.* **2002,** *277*, 31994–32002.
6. Narayanan, S.; Tamura, P. J.; Roth, M. R.; Prasad, P. V.; Welti, R. Wheat Leaf Lipids During Heat Stress: I. High day and Night Temperatures Result in Major Lipid Alterations. *Plant Cell Environ.* **2016,** *39*, 787–803.
7. Djanaguiraman, M.; Boyle, D. L.; Welti, R.; Jagadish, S. V. K.; Prasad, P. V. V. Decreased Photosynthetic Rate Under High Temperature in Wheat Is Due to Lipid Desaturation, Oxidation, Acylation, and Damage of Organelles. *BMC Plant Biol.***2018,** *18*, p.55.
8. Narayanan, S.; Zoong-Lwe, Z. S.; Gandhi, N.; Welti, R.; Fallen, B.; Smith, J. R.; Rustgi, S. Comparative Lipidomic Analysis Reveals Heat Stress Responses of Two Soybean Genotypes Differing in Temperature Sensitivity. *Plants* **2020,** *9*, 457.
9. ZoongLwe, Z.; Sah, S.; Persaud, L.; Li, J.; Gao, W.; Raja Reddy, K.; Narayanan, S. Alterations in the Leaf Lipidome of *Brassica carinata* Under High-Temperature Stress. *BMC Plant Biol.* **2021,** *21*, 404.
10. Narayanan, S.; Prasad, P. V.; Welti, R. Wheat Leaf Lipids During Heat Stress: II. Lipids Experiencing Coordinated Metabolism Are Detected by Analysis of Lipid co-Occurrence. *Plant Cell Environ.* **2016,** *39*, 608–617.
11. Levitt, J. *Responses of Plants to Environmental Stress, Volume 1: Chilling, Freezing, and High Temperature Stresses*; Academic Press: New York, 1980.
12. Zheng, G.; Tian, B.; Zhang, F.; Tao, F.; Li, W. Plant Adaptation to Frequent Alterations Between High and Low Temperatures: Remodelling of Membrane Lipids and Maintenance of Unsaturation Levels. *Plant Cell Environ.* **2011,** *34*, 1431–1442.
13. Narayanan, S.; Membrane Fluidity and Compositional Changes in Response to High Temperature Stress in Wheat. In Physiological, Molecular, and Genetic Perspectives of Wheat Improvement; Wani, S. H., Mohan, A., Singh, G. P., Eds.; Springer International Publishing: Cham, Switzerland, **2021.**
14. Murakami, Y.; Tsuyama, M.; Kobayashi, Y.; Kodama, H.; Iba, K. Trienoic Fatty Acids and Plant Tolerance of High Temperature. *Science* **2000,** *287*, 476–479.
15. von Koskull-Döring, P.; Scharf, K. D.; Nover, L. The Diversity of Plant Heat Stress Transcription Factors. *Trends Plant Sci.* **2007,** *12*, 452–457.
16. Higashi, Y.; Okazaki, Y.; Takano, K.; Myouga, F.; Shinozaki, K.; Knoch, E.; Fukushima, A.; Saito, K. Heat Inducible Lipase1 Remodels Chloroplastic monogalactosyldiacylglycerol by Liberating α-Linolenic Acid in Arabidopsis Leaves Under Heat Stress. *Plant Cell* **2018,** *30*, 1887–1905.
17. Wenk, MR.; Lipidomics: New Tools and Applications. *Cell* **2010,** *143*, 888–895.
18. van Meer, G.; Voelker, D. R.; Feigenson, G. W. Membrane Lipids: Where They Are and How They Behave. *Nat. Rev. Mol. Cell Biol.* **2008,** *9*, 112–124.
19. Nakamura, M. T.; Yudell, B. E.; Loor, J. J.; Regulation of Energy Metabolism by Long-Chain Fatty Acids. *Prog. Lipid Res.* **2014,** *53*, 124–144.

20. Wang, X.; Chapman, K. D. Lipid Signaling in Plants. *Front. Plant Sci.* **2013,** *4*: p. 216.
21. Wallis, J. G.; Browse, J. Mutants of Arabidopsis Reveal Many Roles for Membrane Lipids. *Prog. Lipid Res.* **2002,** *41*, 254–278.
22. Welti, R.; Wang, X. Lipid Species Profiling: A High-Throughput Approach to Identify Lipid Compositional Changes and Determine the Function of Genes Involved in Lipid Metabolism and Signaling. *Curr. Opin. Plant Biol.* **2004,** *7*, 337–344.
23. Tabassum, R.; Ripatti, S. Integrating Lipidomics and Genomics: Emerging Tools to Understand Cardiovascular Diseases. *Cell. Mol. Life Sci. CMLS* **2021,** *78*, 2565–2584.
24. Reszczy´nska, E.; Hanaka, A. Lipids Composition in Plant Membranes. *Cell Biochem. Biophys.* **2020,** *78*, 401–414.
25. Kunst, L.; Browse, J.; Somerville, C. Altered Chloroplast Structure and Function in a Mutant of Arabidopsis Deficient in Plastid Glycerol-3-Phosphate Acyltransferase Activity. *Plant Physiol.* **1989,** *90*, 846–853.
26. Carman, G. M.; Henry, S. A. Phosphatidic Acid Plays a Central Role in the Transcriptional Regulation of Glycerophospholipid Synthesis in *Saccharomyces cerevisiae*. *J. Biol. Chem.* **2007**, *282*, 37293–37297.
27. Duchêne, S.; Siegenthaler, P. A. Do Glycerolipids Display Lateral Heterogeneity in the Thylakoid Membrane? *Lipids* **2000,** *35*, 739–744.
28. Jones, M. R. Lipids in Photosynthetic Reaction Centres: Structural Roles and Functional Holes. *Prog. Lipid Res.* **2007,** *46*, 56–87.
29. Nelson, D. L.; Cox, M. M. *Lehninger Principles of Biochemistry*, 5th ed.; Macmillan: London, 2014.
30. Michaelson, L. V.; Napier, J. A.; Molino, D.; Faure, J. D. Plant Sphingolipids: Their Importance in Cellular Organization and Adaption. *Biochim. Biophys. Acta* **2016,** *1861*, 1329–1335.
31. MamodeCassim, A.; Gouguet, P.; Gronnier, J.; Laurent, N.; Germain, V.; Grison, M.; Boutté, Y.; Gerbeau-Pissot, P.; Simon-Plas, F.; Mongrand, S. Plant Lipids: Key Players of Plasma Membrane Organization and Function. *Prog. Lipid Res.* **2019,** *73*, 1–27.
32. Valitova, J. N.; Sulkarnayeva, A. G.; Minibayeva, F. V. Plant Sterols: Diversity, Biosynthesis, and Physiological Functions. *Biochemistry* **2016,** *81*, 819–834.
33. Pelillo, M.; Iafelice, G.; Marconi, E.; FiorenzaCaboni, M.; Identification of Plant Sterols in Hexaploid and Tetraploid Wheats Using Gas Chromatography with Mass Spectrometry. *Rapid Commun. Mass Spectrom.* **2003,** *17*, 2245–2252.
34. Hartmann, M. A. Plant Sterols and the Membrane Environment. *Trends Plant Sci.* **1998,** *3*, 170–175.
35. Watson, A. D.; Thematic Review Series: Systems Biology Approaches to Metabolic and Cardiovascular Disorders—Lipidomics: A Global Approach to Lipid Analysis in Biological Systems. *J. Lipid Res.* **2006,** *47*, 2101–2111.
36. Navarro-Reig, M.; Tauler, R.; Iriondo-Frias, G.; Jaumot, J. Untargeted Lipidomic Evaluation of Hydric and Heat Stresses on Rice Growth. *J. Chromatogr. B Anal. Technol. Biomed. Life Sci.* **2019,** *1104*, 148–156.
37. Tumanov, S.; Kamphorst, J. J. Recent Advances in Expanding the Coverage of the Lipidome. *Curr. Opin. Biotechnol.* **2017,** *43*, 127–133.

38. Higashi, Y.; Saito, K. Lipidomic Studies of Membrane Glycerolipids in Plant Leaves Under Heat Stress. *Prog. Lipid Res.* **2019,** *75*, 100990.
39. Züllig, T.; Trötzmüller, M.; Köfeler, H. C. Lipidomics from Sample Preparation to Data Analysis: A Primer. *Anal. Bioanal. Chem.* **2020,** *412*, 2191–2209.
40. Han, X.; Yang, K.; Gross, R. W.; Multi-Dimensional Mass Spectrometry-Based Shotgun Lipidomics and Novel Strategies for Lipidomic Analyses. *Mass Spectrom. Rev.* **2012,** *31*, 134–178.
41. Wang, M.; Wang, C.; Han, X. Selection of Internal Standards for Accurate Quantification of Complex Lipid Species in Biological Extracts by Electrospray Ionization Mass Spectrometry—What, How and Why? *Mass Spectrom. Rev.* **2017,** *36*, 693–714.
42. Khoury, S.; Canlet, C.; Lacroix, M. Z.; Berdeaux, O.; Jouhet, J.; Bertrand-Michel, J.; Quantification of Lipids: Model, Reality, and Compromise. *Biomolecules* **2018,** *8*, p. 174.
43. Ivanova, P. T.; Milne, S. B.; Myers, D. S.; Brown, H. A. Lipidomics: A Mass Spectrometry Based Systems Level Analysis of Cellular Lipids. *Curr. Opin. Chem. Biol.* **2009,** *13*, 526–531.
44. Shiva, S.; Vu, H. S.; Roth, M. R.; Zhou, Z.; Marepally, S. R.; Nune, D. S.; Lushington, G. H.; Visvanathan, M.; Welti, R. Lipidomic Analysis of Plant Membrane Lipids by Direct Infusion Tandem Mass Spectrometry. *Methods Mol. Biol.* **2013,** *1009*, 79–91.
45. McDonald, J. G.; Smith, D. D.; Stiles, A. R.; Russell, D. W. A Comprehensive Method for Extraction and Quantitative Analysis of Sterols and Secosteroids from Human Plasma. *J. Lipid Res.* **2012,** *53*, 1399–1409.
46. Shulaev, V.; Chapman, K. D. Plant Lipidomics at the Crossroads: From Technology to Biology Driven Science. *Biochim. Biophys. Acta. Mol. Cell Biol. Lipids* **2017,** *1862*, 786–791.
47. Cajka, T.; Fiehn, O. Comprehensive Analysis of Lipids in Biological Systems by Liquid Chromatography-Mass Spectrometry. *Trends Anal. Chem.* **2014,** *61*, 192–206.
48. Wang, M.; Wang, C.; Han, R. H.; Han, X. Novel Advances in Shotgun Lipidomics for Biology and Medicine. *Prog. Lipid Res.* **2016,** *61*, 83–108.
49. Welti, R.; Shah, J.; Li, W.; Li, M.; Chen, J.; Burke, J. J.; Fauconnier, M. L.; Chapman, K.; Chye, M. L.; Wang, X. Plant Lipidomics: Discerning Biological Function by Profiling Plant Complex Lipids Using Mass Spectrometry. *Front. Biosci. J. Virtual Libr.* **2007,** *12*, 2494–2506.
50. Brügger, B.; Erben, G.; Sandhoff, R.; Wieland, F. T.; Lehmann, W. D. Quantitative Analysis of Biological Membrane Lipids at the Low Picomole Level by Nano-Electrospray Ionization Tandem Mass Spectrometry. *Proc. Nat. Acad. Sci. USA* **1997,** *94*, 2339–2344.
51. Welti, R.; Wang, X.; Williams, T. D. Electrospray Ionization Tandem Mass Spectrometry Scan Modes for Plant Chloroplast Lipids. *Anal. Biochem.* **2003,** *314*, 149–152.
52. Devaiah, S. P.; Roth, M. R.; Baughman, E.; Li, M.; Tamura, P.; Jeannotte, R.; Welti, R.; Wang, X. Quantitative Profiling of Polar Glycerolipid Species from Organs of Wild-Type Arabidopsis and a PHOSPHOLIPASE Dα1 Knockout Mutant. *Phytochemical* **2006,** *67*, 1907–1924.

53. Han, X.; Yang, K.; Gross, R. W. Microfluidics-Based Electrospray Ionization Enhances the Intrasource Separation of Lipid Classes and Extends Identification of Individual Molecular Species Through Multi-Dimensional Mass Spectrometry: Development of an Automated High-Throughput Platform for Shotgun Lipidomics. *Rapid Commun. Mass Spectrom.* **2008,** *22*, 2115–2124.
54. Lam, S. M.; Tian, H.; Shui, G. Lipidomics, En Route to Accurate Quantitation. *Biochim. Biophys. Acta Mol. Cell Biol. Lipids* **2017,** *1862*, 752–761.
55. Dowhan, W. Molecular Genetic Approaches to Defining Lipid Function. *J. Lipid Res.* **2009,** *50*, S305–S310.
56. Sanchez, J.; Mangat, P. K.; Angeles-Shim, R. B. Weathering the Cold: Modifying Membrane and Storage Fatty Acid Composition of Seeds to Improve Cold Germination Ability in Upland Cotton (*Gossypiumhirsutum* L.). *Agronomy* **2019,** *9*, 684.
57. Conde, A.; Chaves, M. M.; Gerós, H. Membrane Transport, Sensing and Signaling in Plant Adaptation to Environmental Stress. *Plant Cell Physiol.* **2011,** *52*, 1583–1602.
58. ur Rahman, H.; Malik, S. A.; Saleem, M. Heat Tolerance of Upland Cotton During the Fruiting Stage Evaluated Using Cellular Membrane Thermostability. *Field Crops Res.* **2004,** *85*, 149–158.
59. Jouhet, J. Importance of the Hexagonal Lipid Phase in Biological Membrane Organization. *Front. Plant Sci.* **2013,** *4*, 494.
60. Sonnino, S.; Prinetti, A. Membrane Domains and the "Lipid Raft" Concept. *Curr. Med. Chem.* **2013,** *20*, 4–21.
61. Edidin, M. Lipids on the Frontier: A Century of Cell-Membrane Bilayers. *Nat. Rev. Mol. Cell Biol.* **2003,** *4*, 414–418.
62. Larkindale, J.; Huang, B. Changes of Lipid Composition and Saturation Level in Leaves and Roots for Heat-Stressed and Heat Acclimated Creeping Bentgrass (*Agrostis stolonifera*). *Environ. Exp. Bot.* **2004,** *51*, 57–67.
63. Quinn, P. J. A Lipid-Phase Separation Model of Low-Temperature Damage to Biological Membranes. *Cryobiology* **1985,** *22*, 128–146.
64. Liu, X.; Huang, B. Heat Stress Injury in Relation to Membrane Lipid Peroxidation in Creeping Bentgrass. *Crop Sci.* **2000,** *40*, 503–510.
65. Yamamoto, Y.; Aminaka, R.; Yoshioka, M.; Khatoon, M.; Komayama, K.; Takenaka, D.; Yamashita, A.; Nijo, N.; Inagawa, K.; Morita, N. et al. Quality Control of Photosystem II: Impact of Light and Heat Stresses. *Photosynth. Res.* **2008,** *98*, 589–608.
66. Yamamoto, Y.; Kai, S.; Ohnishi, A.; Tsumura, N.; Ishikawa, T.; Hori, H.; Morita, N.; Ishikawa, Y.; Quality control of PSII: Behavior of PSII in the highly crowded grana thylakoids under excessive light. *Plant Cell Physiol.* **2014,** *55*, 1206–1215.
67. Hasanuzzaman, M.; Nahar, K.; Alam, M. M.; Roychowdhury, R.; Fujita, M. Physiological, Biochemical, and Molecular Mechanisms of Heat Stress Tolerance in Plants. *Int. J. Mol. Sci.* **2013,** *14*, 9643–9684.
68. Niu, Y.; Xiang, Y. An Overview of Biomembrane Functions in Plant Responses to High-Temperature Stress. *Front. Plant Sci.* **2018,** *9*, 915.
69. Qaseem, M. F.; Qureshi, R.; Shaheen, H. Effects of Pre-Anthesis Drought, Heat and Their Combination on the Growth, Yield and Physiology of Diverse Wheat

(Triticumaestivum L.) Genotypes Varying in Sensitivity to Heat and Drought Stress. *Sci. Rep.* **2019,** *9*, 6955.
70. Li, G.; Chen, T.; Feng, B.; Peng, S.; Tao, L.; Fu, G. Respiration, Rather Than Photosynthesis, Determines Rice Yield Loss Under Moderate High-Temperature Conditions. *Front. Plant Sci.* **2021,** *12*, 678653.
71. Prasad, P. V. V.; Djanaguiraman, M. High Night Temperature Decreases Leaf Photosynthesis and Pollen Function in Grain Sorghum. *Funct. Plant Biol. FPB* **2011,** *38*, 993–1003.
72. Djanaguiraman, M.; Prasad, P. V.; Boyle, D.; Schapaugh, W.; High-Temperature Stress and Soybean Leaves: Leaf Anatomy and Photosynthesis. *Crop Sci.* **2011,** *51*, 2125–2131.
73. Hussain, H. A.; Men, S.; Hussain, S.; Chen, Y.; Ali, S.; Zhang, S.; Zhang, K.; Li, Y.; Xu, Q.; Liao, C. et al. Interactive Effects of Drought and Heat Stresses on Morpho-Physiological Attributes, Yield, Nutrient Uptake and Oxidative Status in Maize Hybrids. *Sci. Rep.* **2019,** *9*, 3890.
74. Sarwar, M.; Saleem, M. F.; Ullah, N.; Ali, S.; Rizwan, M.; Shahid, M. R.; Alyemeni, M. N.; Alamri, S. A.; Ahmad, P. Role of Mineral Nutrition in Alleviation of Heat Stress in Cotton Plants Grown in Glasshouse and Field Conditions. *Sci. Rep.* **2019,** *9*, 13022.
75. Webb, M. S.; Green, B. R. Biochemical and Biophysical Properties of Thylakoid Acyl Lipids. *Biochim. Biophys. Acta Bioenerg.* **1991,** *1060*, 133–158.
76. Falcone, D. L.; Ogas, J. P.; Somerville, C. R. Regulation of Membrane Fatty Acid Composition by Temperature in Mutants of Arabidopsis with Alterations in Membrane Lipid Composition. *BMC Plant Biol.* **2004,** *4*, 17.
77. Xu, L.; Pan, R.; Zhou, M.; Xu, Y.; Zhang, W. Lipid Remodelling Plays an Important Role in Wheat (Triticumaestivum) Hypoxia Stress. *Funct. Plant Biol. FPB* **2019,** *47*, 58–66.
78. Rustgi, S.; Kakati, J. P.; Jones, Z. T.; ZoongLwe, Z. S.; Narayanan, S. Heat Tolerance as a Function of Membrane Lipid Remodeling in the Major US Oilseed Crops (Soybean and Peanut). *J. Plant Biochem. Biotechnol.* **2021,** *30*, 652–667.
79. Abdelrahman, M.; Ishii, T.; El-Sayed, M.; Tran, L. S. P. Heat Sensing and Lipid Reprograming as a Signaling Switch for Heat Stress Responses in Wheat. *Plant Cell Physiol.* **2020,** *61*, 1399–1407.
80. Mueller, S. P.; Krause, D. M.; Mueller, M. J.; Fekete, A. Accumulation of Extra-Chloroplastictriacylglycerols in Arabidopsis Seedlings During Heat Acclimation. *J. Exp. Bot.* **2015,** *66*, 4517–4526.
81. Mueller, S. P.; Unger, M.; Guender, L.; Fekete, A.; Mueller, M. J.; Phospholipid: Diacylglycerol Acyltransferase-Mediated Triacylglyerol Synthesis Augments Basal Thermotolerance. *Plant Physiol.* **2017,** *175*, 486–497.
82. Li, Q.; Zheng, Q.; Shen, W.; Cram, D.; Fowler, D. B.; Wei, Y.; Zou, J. Understanding the Biochemical Basis of Temperature-Induced Lipid Pathway Adjustments in Plants. *Plant Cell* **2015,** *27*, 86–103.
83. Jia, Y.; Li, W. Characterisation of Lipid Changes in Ethylene-Promoted Senescence and Its Retardation by Suppression of Phospholipase Dδ in Arabidopsis Leaves. *Front. Plant Sci.* **2015,** *6*, 1045.

84. Muramatsu, K.; Masumizu, T.; Maitani, Y.; Hwang, S. H.; Kohno, M.; Takayama, K.; Nagai, T. Electron Spin Resonance Studies of Dipalmitoylphosphatidylcholine Liposomes Containing Soybean-Derived Sterylglucoside. *Chem. Pharm. Bull.* **2000,** *48*, 610–613.
85. Djanaguiraman, M.; Narayanan, S.; Erdayani, E.; Prasad, P. Effects of High Temperature Stress During Anthesis and Grain Filling Periods on Photosynthesis, Lipids and Grain Yield in Wheat. *BMC Plant Biol.* **2020,** *20*, 1–12.
86. ZoongLwe, Z. S.; Welti, R.; Anco, D.; Naveed, S.; Rustgi, S.; Narayanan, S. Heat Stress Elicits Remodeling in the Anther Lipidome of Peanut. *Sci. Rep.* **2020,** *10*:p. 22163.
87. Marla, S. R.; Shiva, S.; Welti, R.; Liu, S.; Burke, J. J.; Morris, G. P.; Comparative Transcriptome and Lipidome Analyses Reveal Molecular Chilling Responses in Chilling-Tolerant Sorghums. *Plant Genome* **2017,** *10*.
88. Maulana, F.; Ayalew, H.; Anderson, J. D.; Kumssa, T. T.; Huang, W.; Ma, X. F. Genome-Wide Association Mapping of Seedling Heat Tolerance in Winter Wheat. *Front. Plant Sci.* **2018,** *9*, 1272.
89. Shikha, K.; Shahi, J. P.; Vinayan, M. T.; Zaidi, P. H.; Singh, A. K.; Sinha, B. Genome-Wide Association Mapping in Maize: Status and Prospects. *3 Biotech* **2021,** *11*, 244.
90. Bita, C. E.; Gerats, T. Plant Tolerance to High Temperature in a Changing Environment: Scientific Fundamentals and Production of Heat Stress-Tolerant Crops. *Front. Plant Sci.* **2013,** *4*, 273.
91. Cossani, C. M.; Reynolds, M. P. Physiological Traits for Improving Heat Tolerance in Wheat. *Plant Physiol.* **2012,** *160*, 1710–1718.
92. Leung, H. Stressed Genomics—Bringing Relief to Rice Fields. *Curr. Opin. Plant Biol.* **2008,** *11*, 201–208.
93. Chen, J.; Xue, M.; Liu, H.; Fernie, A. R.; Chen, W. Exploring the Genic Resources Underlying Metabolites Through mGWAS and mQTL in Wheat: From Large-Scale Gene Identification and Pathway Elucidation to Crop Improvement. *Plant Commun.* **2021,** *2*, 100216.
94. Liang, X.; Liu, S.; Wang, T.; Li, F.; Cheng, J.; Lai, J.; Qin, F.; Li, Z.; Wang, X.; Jiang, C. Metabolomics-Driven Gene Mining and Genetic Improvement of Tolerance to Salt-Induced Osmotic Stress in Maize. *New Phytol.* **2021,** *230*, 2355–2370.
95. Reynolds, M.; Trethowan, R. Physiological Interventions in Breeding for Adaptation to Abiotic Stress. In: *Scale and Complexity in Plant Systems Research, Gene–Plant–Crop Relations*; Spiertz, J. H. J., Struik, P. C., Van Laar, H. H., Eds.; UR Frontis Series: Wageningen, The Netherlands, 2007; pp 127–144.
96. Luo, J. Metabolite-Based Genome-Wide Association Studies in Plants. *Curr. Opin. Plant Biol.* **2015,** *24*, 31–38.
97. Raza, A. Metabolomics: A Systems Biology Approach for Enhancing Heat Stress Tolerance in Plants. *Plant Cell Rep.* **2022,** *41*, 741–763.
98. Chen, D.; Wang, S.; Qi, L.; Yin, L.; Deng, X. Galactolipidremodeling Is Involved in Drought-Induced Leaf Senescence in Maize. *Environ. Exp. Bot.* **2018,** *150*, 57–68.
99. Yang, W.; Feng, H.; Zhang, X.; Zhang, J.; Doonan, J. H.; Batchelor, W. D.; Xiong, L.; Yan, J. Crop Phenomics and High-Throughput Phenotyping: Past Decades, Current Challenges, and Future Perspectives. *Mol. Plant.* **2020,** *13*, 187–214.

100. Fiehn, O. Metabolomics—The Link Between Genotypes and Phenotypes. *Plant Mol. Biol.* **2002,** *48*, 155–171.
101. Sue, M.; Yamazaki, K.; Yajima, S.; Nomura, T.; Matsukawa, T.; Iwamura, H.; Miyamoto, T. Molecular and Structural Characterization of Hexameric β-d-Glucosidases in Wheat and Rye. *Plant Physiol.* **2006,** *141*, 1237–1247.
102. Saito, K.; Hirai, M. Y.; Yonekura-Sakakibara, K. Decoding Genes with Coexpression Networks and Metabolomics—'Majority Report by Precogs'. *Trends Plant Sci.* **2008,** *13*, 36–43.
103. de Abreu e Lima, F.; Li, K.; Wen, W.; Yan, J.; Nikoloski, Z.; Willmitzer, L.; Brotman, Y. Unraveling Lipid Metabolism in Maize with Time-Resolved Multi-Omics Data. *Plant J.* **2018,** *93*, 1102–1115.
104. Beleggia, R.; Rau, D.; Laido, G.; Platani, C.; Nigro, F.; Fragasso, M.; De Vita, P.; Scossa, F.; Fernie, A. R.; Nikoloski, Z. et al. Evolutionary Metabolomics Reveals Domestication-Associated Changes in Tetraploid Wheat Kernels. *Mol. Biol.* **2016,** *33*, 1740–1753.
105. Riedelsheimer, C.; Lisec, J.; Czedik-Eysenberg, A.; Sulpice, R.; Flis, A.; Grieder, C.; Altmann, T.; Stitt, M.; Willmitzer, L.; Melchinger, A. E. Genome-Wide Association Mapping of Leaf Metabolic Profiles for Dissecting Complex Traits in Maize. *Proc. Natl. Acad. Sci. USA* **2012,** *109*, 8872–8877.
106. Chen, W.; Gao, Y.; Xie, W.; Gong, L.; Lu, K.; Wang, W.; Li, Y.; Liu, X.; Zhang, H.; Dong, H. Genome-Wide Association Analyses Provide Genetic and Biochemical Insights into Natural Variation in Rice Metabolism. *Nat. Genet.* **2014,** *46*, 714–721.
107. Wen, W.; Li, K.; Alseekh, S.; Omranian, N.; Zhao, L.; Zhou, Y.; Xiao, Y.; Jin, M.; Yang, N.; Liu, H. et al. Genetic Determinants of the Network of Primary Metabolism and Their Relationships to Plant Performance in a Maize Recombinant Inbred Line Population. *Plant Cell* **2015,** *27*, 1839–1856.
108. Chen, W.; Wang, W.; Peng, M.; Gong, L.; Gao, Y.; Wan, J.; Wang, S.; Shi, L.; Zhou, B.; Li, Z. Comparative and Parallel Genome-Wide Association Studies for Metabolic and Agronomic Traits in Cereals. *Nat. Commun.* **2016,** *7*, 1–10.
109. Templer, S. E.; Ammon, A.; Pscheidt, D.; Ciobotea, O.; Schuy, C.; McCollum, C.; Sonnewald, U.; Hanemann, A.; Förster, J.; Ordon, F. Metabolite Profiling of Barley Flag Leaves Under Drought and Combined Heat and Drought Stress Reveals Metabolic QTLs for Metabolites Associated with Antioxidant Defense. *J. Exp. Bot.* **2017,** *68*, 1697–1713.
110. Schauer, N.; Semel, Y.; Roessner, U.; Gur, A.; Balbo, I.; Carrari, F.; Pleban, T.; Perez-Melis, A.; Bruedigam, C.; Kopka, J. et al. Comprehensive Metabolic Profiling and Phenotyping of Interspecific Introgression Lines for Tomato Improvement. *Nat. Biotechnol.* **2006,** *24*, 447–454.
111. Wei, W.; Li, S.; Wang, Y.; Wang, B.; Fan, G.; Zeng, Q.; Zhao, F.; Xu, C.; Zhang, X.; Tang, T. et al. Metabolome-Based Genome-Wide Association Study Provides Genetic Insights into the Natural Variation of Foxtail Millet. *Front. Plant Sci.* **2021,** *12*, 665530.
112. Flint-Garcia, S. A.; Thornsberry, J. M.; Buckler, E. S. Structure of Linkage Disequilibrium in Plants. *Annu. Rev. Plant Biol.* **2003,** *54*, 357–374.
113. Mauricio, R. Mapping Quantitative Trait Loci in Plants: Uses and Caveats for Evolutionary Biology. *Nat. Rev. Genet.* **2001,** *2*, 370–381.

114. Chen, J.; Hu, X.; Shi, T.; Yin, H.; Sun, D.; Hao, Y.; Xia, X.; Luo, J.; Fernie, A. R.; He, Z. et al. Metabolite-Based Genome-Wide Association Study Enables Dissection of the Flavonoid Decoration Pathway of Wheat Kernels. *Plant Biotechnol. J.* **2020,** *18*, 1722–1735.
115. Zhu, G.; Wang, S.; Huang, Z.; Zhang, S.; Liao, Q.; Zhang, C.; Lin, T.; Qin, M.; Peng, M.; Yang, C. et al. Rewiring of the Fruit Metabolome in Tomato Breeding. *Cell* **2018,** *172*, 249–261.e212.

PART III

REPRODUCTIVE GROWTH AND NIGHT WARMING

CHAPTER 8

Effect of Night-time Warming on Reproductive Growth of Plants

SAURABH PANDEY[1], SWATI PURANIK[2], and PRANAV PANKAJ SAHU[2]

[1]*Department of Agriculture, Guru Nanak Dev University, Amritsar, Punjab, India*

[2]*Global Change Research Institute of the Czech Academy of Sciences, Bělidla, Brno, Czechia*

ABSTRACT

Global climate change has resulted in incremental temperature variations. This is a major concern for proper reproductive growth in plants. Increasingly alarming episodes of poor pollen development, floret fertility, and significant yield losses have become widespread across various species. Current understanding of heat stress-induced modulations in pollen morphology and its biochemical composition, stigma, reproductive phase alterations, and sensitivity indicates the direct correlation between seed set and yield loss in different crops. Heat stress research in crop plants has been majorly focused on temperature variations incurred during the daytime. However, diurnal temperature variations with night-time warming result in an emerging threat to pollen development and subsequent productivity. An interesting way of coping with this stress, by escaping flowering at the sensitive phase and early morning flowering, has been tapped in the wild species of plants. Understanding the mechanism of this escape and underlying molecular and biochemical changes will help us to breed improved climate-smart cultivars which are adapted to tolerate heat stress

Impact of High Night Temperature on Plant Biology: Toward Sustainable Plant Adaptation to Climate Change. Rajesh Kumar Singhal, Raju Bheemanahalli, Saurabh Pandey, and Pratibha M. D. (Eds.)

and maintain productivity. This chapter emphasizes the importance of reproductive stages for crop yield along with how heat stress, mainly night temperature, affects pollen development, and finally discusses the resources and mechanisms to avoid these losses.

8.1 INTRODUCTION

Global warming and associated climate change have increased global temperatures posing a severe threat to crop production and food productivity. Unfavorable temperatures, especially during reproductive stages, compromise the plant biomass, seed set, seed characteristics, and, ultimately, the yield. Different crops, including cereals, oilseeds, pulses, millets, etc., behave differently at high-temperature exposure depending upon the duration and the intensity of stress.[1,2] This is reflected by their harvest index changes under heat stress and optimum temperature threshold for crops that could present an alternative for particular geographical regions. Under controlled temperature conditions, the damaging ranges of temperatures for reproductive stages have been quantified to be in the range of 30–40°C. For example, temperature coinciding at the reproductive stage for cereals (30–38°C), oilseeds (35–36°C), pulses (32–40°C), and millets (40°C) reduced the crop yield significantly.[3] Another factor that significantly prolongs a plant's heat exposure is the narrowing of diurnal temperature amplitude as reflected by the rapid increase in warmer nights.[4] Negative impacts of night-time warming on grain yield and quality have been reported for various crops.[5–13] Night-time warming drastically affects plant photosynthesis and respiration to modulate the internal carbohydrate pool, resulting in reduced biomass.[5,7,11]

A significant factor determining seed set and productivity is the growth and development of male reproductive organs such as pollens or anthers. Prolonged reproductive stage heat stress leads to morphophysiological changes during pollen development and flowering which are the most susceptible reproductive phases against temperature stress.[14–16] Under temperature stress, these processes are delayed or extended for 2 or 3 weeks based on the species. Further, reduced pollen germination percentage on the stigmatic surfaces and its viability loss leads to infertility in different crops.[17–20] This effect has been widely observed for pollen tube growth and development and pollination in wheat, rice, chickpea, cotton, and other crops.[21–26] However, the information is limited to pollen fertility, and more attention is needed to understand the impact on female reproductive organs during heat stress. Mechanistic insights for pollen and stigma morphology, biochemical composition, membrane damage, and molecular mechanisms during heat onset remain unexplored for major crops. Therefore, exploring the developmental processes at reproductive stages can allow for curating delicate genetic mechanisms and related molecular interventions to cope with the detrimental effects of temperature stress.

Crop improvement efforts for the longest time focused on the yield increase in the major crops, resulting in the loss of stress-responsive genes and alleles in popular cultivars, making them vulnerable to environmental stress. To manage the yield losses pertaining to high-temperature exposure at reproductive stages, there is a need to identify stress avoidance mechanisms that the plants adopt. For example, early morning flowering has been identified as a heat escape mechanism in rice.[27] Similarly, transpiration cooling and resilience during reproductive stress also help the plant to escape high temperatures and yield losses.[22,28] Quantification of such traits for genetic gains remains a challenge to harness them in a breeding program. Exploring the genetic diversity for resilience traits in wild species could provide a mechanism for sustainable genetic gains for heat stress, for example, in wheat crops.[29–30] This chapter explores different factors leading to reproductive stage susceptibility and how day and night high temperatures affect cellular and molecular mechanisms. Various sources of tolerance to cope with this stress, including related wild alleles will be discussed. Overall, this chapter will enhance our knowledge about understanding night temperature stress on reproductive stages and their possible solutions.

8.2 REPRODUCTIVE STAGE SENSITIVITY: EFFECT OF TIME, TEMPERATURE AND CANOPY POSITION

Plant reproductive processes comprise pollen development, ovary development, and fertilization, which are extremely sensitive to high temperatures, particularly for field crops. For example, in cereals such as rice, the time for completing their flowering was found to increase from 5–6 days to 14 days.[31] Similarly, in oilseed crops, owing to their indeterminate nature, the flowering time could increase by up to 30 days, compromising the yield and quality. Scientists have identified that major field crops narrow their reproductive stage sensitivity time range, focusing more on pollen development and sterility. Exploring and identifying the sensitivity window precisely is crucial to devise a practical approach for minimizing the damage. However, differences in flowering patterns and crop growth habits made this determination difficult. Nevertheless, a typical period of sensitivity of 5–9 days before anthesis and during gametogenesis was identified in rice, wheat, peanut, and sorghum.[2,32–34] These studies indicated that the stress-affected open flowers exposed the floral parts to high temperatures and resulting in the loss of viability.[32,35,36]

Crops have been found to vary extensively in the threshold temperature, and any extension of this or longer duration exposure to high temperatures negatively impacts the vegetative and grain-filling stages, ultimately affecting crop yield. Recent high temperatures during February and March in the Asian subcontinent which coincided with the reproductive phase of wheat crop resulted in significant yield losses.[37] An increase in daily mean temperature above 16°C significantly dropped the harvest index and yield in wheat. Temperatures around 30°C or more result in complete yield loss in wheat, whereas for other cereals, this threshold is around 40°C as seen in pearl millet (41°C), peanut (40°C), soybean (39°C), sorghum (38°C), rice (35°C), and wheat (31°C).[3]

Besides these threshold temperatures, the exposure of floral organs above the canopy level decreases the grain yield significantly in wheat and rice. Branching in peanuts and soybeans safeguards the flowers during reproductive phases against direct exposure. Also, profuse canopy growth provides transpiration cooling to lower the temperatures for floral organs and helps in successful fertilization, earlier described as heat avoidance in rice and needs exploration in other crops.[38,39] Most heat avoidance work was done in rice to investigate the escape mechanism induced during high-temperature exposures.[22,27,40] Like rice, other cereals such as pearl millet and sorghum are more resilient than the wheat crop despite floral exposure to heat identical to wheat. Therefore, it will be interesting to explore the mechanism and candidate identification in these crops. This exploration will further add to the targeted response by altering the flowering pattern, timing, and architecture to effectively modulate the field crop microclimate and enhance our field crop's resilience.

8.3 CELLULAR PROCESSES AFFECTED BY HEAT STRESS AT THE REPRODUCTIVE STAGE

Several morphological, structural, and biochemical changes are induced due to high-temperature exposure that alter the lipid composition and carbohydrate partitioning and affect the fertilization and development of pods or seeds. In a recent study, high-temperature stress was found to be affecting pollen–pistil interaction and hindered fertilization due to damage to style, stigma, and ovary.[3] The synchronous development of male and female reproductive organs is responsible for seed set in field crops which involves a sequence of signaling cascades to complete fertilization and embryo development. For example, heat stress modified the maturation time of male and female reproductive organs in wheat[21] and ovule numbers, and fertility in common beans.[41] It was found to delay embryo sac development in mustard,[20] shorten the stigma receptivity in chickpeas,[23] and collapse and flattened the pollen in soybean, barley, and other crops.[42–44] Structural components required during pollen tube growth, like microtubules, are also disrupted by exposure to high temperatures and modulate the signaling and vesicular transport, impairing fertilization.[45]

Biochemical changes, such as increased mitochondrial numbers at the early active microspore stages, are essential for meeting the energy demand for tapetum development.[46,47] This number increase leads to a higher accumulation of reactive oxygen species (ROS) in reproductive organs. Higher mitochondrial numbers consequently increase the pollen's respiration and result in the depletion of food reserves, which could be a potential factor for reproductive stage sensitivity during high-temperature stress. The respiration and ROS production increase also imbalances the ROS scavenging enzymes to severe high temperature-related damages. In a short exposure to heat stress in wheat and sorghum, increased anther and pollen ROS resulted in increased heat sensitivity.[17,48] It also damages membrane lipids affecting their permeability and overall functions. ROS accumulation in reproductive tissues also tends to induce misfolding of nascent proteins.[49]

A critical factor for pollen viability is proper sucrose biosynthesis and hydrolysis during the development of the reproductive organs. High-temperature stress in sorghum impaired the sucrose biosynthesis in the pollen grain, thus resulting in reduced viability of pollens.[50] It is shown that long-term exposure to higher temperatures results in reduced expression of sugar transporter genes, sugar enzymes such as invertase as well as starch biosynthesis genes in sorghum crops during gametangia development.[50,51] This decrease in gene expression ultimately leads to spikelet fertility and fertilization failure. Identifying transporters and candidate genes for sugar transporters at the sensitivity stage could be a viable option to mitigate high-temperature-induced damages during day and night for field crops.

Lipid profiling or lipidome analysis of the pollen grain could provide insights into its composition and viability. It is observed that the thickness of the exine wall and the tapetal wall is affected by the high temperature, which determines its degradation.[7,17,52] The pollen viability depends upon the exine wall's lipid composition and determines the fertilization fate with stigmatic surface interactions and signaling.[53] Specific lipids, such as phosphatidic lipids, determine the amount of unsaturated fatty acids in the pollen membrane. Lipid profiling in sorghum showed that high-temperature stress changed its phosphatidic content to compromise the membrane integrity.[9] Heat stress-induced lipid composition changes in pollen modify the stigma receptivity, increase style desiccation, and damage the ovary, which ultimately reduces the number of pollen grains and fertilization success rate. In pearl millet, stigma, and other female reproductive parts were shown to be more affected in comparison with male reproductive parts by the high-temperature stress.[54] It is interesting to understand the susceptibility of female reproductive portions and their interactions with pollen grains which needs comprehensive analysis in major field crops.

8.4 EFFECT OF NIGHT-TIME WARMING DURING REPRODUCTIVE STAGES

Research on the effects of heat stress during the reproductive stage has mostly focused on maximum day temperature and its direct correlation with reproductive failures and grain yield losses in field crops. However, night-time temperature warming could impact the crop sensitivity window during the reproductive phase. Recent climate change reports suggest a significant increase in night-time temperatures, leading to season-long high-temperature stress during floral development and resultant productivity loss.[4,13] Controlled condition studies for increased night-time temperatures suggested significant sterility and yield losses in different crops.[9,10,55,56] Interestingly, field-level studies for high night-time temperatures show a marginal effect for spikelet fertility but a significant reduction in plant biomass due to increased night respiration leading to carbohydrate losses from all plant parts, including panicle and yield loss.[5,11,57] For example, growing wheat and barley in increased night-time temperatures (more than 6°C from ambient temperature) during the post-anthesis to grain filling stage resulted in a 6–7% reduction in grain yield for each °C temperature increase.[6] This decrease was majorly attributed to spikes reduction, resulting in lower seed set and grain yield. One major complication in these studies is the high cost of controlled and field-level experiments to precisely measure the impact of daytime and night-time temperatures on field crops. Accurate prediction of different stages of floral organs' night and day temperature susceptibility, and the underlying physiological and molecular mechanisms requires techniques such as high throughput phenotyping. Additionally, measuring reproductive stage susceptibility in other crops will help to identify possible mechanisms of heat escape and underlying factors and genes that could be explored further for climate-resilient crops in projected high-temperature episodes. Therefore, protocols for specific crop groups with detailed response studies will pave the path for future studies.

8.5 SOURCES FOR REPRODUCTIVE STAGE HIGH-TEMPERATURE TOLERANCE

Identifying traits/factors/genes/alleles for tolerance at the reproductive stage will help breed high-temperature resilient crops. Most breeding programs for developing popular cultivars were focused on yield increase and thus lost crucial stress-tolerance-related genes during the process. Therefore, it is high time to tap the diversity among the wild-type species, accessions, and germplasm to overcome this bottleneck in the current reproductive stage stress tolerance breeding program among major cereals. For example, time of day time flowering in rice has been identified as a stress-resilient trait, and *Oryza officinalis* was shown to avoid high temperatures by early morning flowering.[27,58] This was also incorporated into the breeding programs for popular cultivars utilizing molecular breeding.[27,58] High temperature in nature is often associated with water deficit conditions in the field, resulting in increased floral tissue temperatures and negatively affecting their growth and development.[59] This co-occurrence of both stresses was more prominent when the plant favored the early noon flowering in tropical climates.[60] Therefore, breeding sensitive cereals for early morning flowering could be a potential trait of interest focused on candidate identification and discovering the underlying mechanisms.

Diversity was also observed in wild wheat species for heat stress tolerance. For example, *Aegilops speltoides* and *A. geniculata* were more efficient in avoiding high-temperature stress than popular hexaploid wheat cultivars.[29] Pradhan et al.[30] found that wheat breeding programs involving chromosomal substitution lines and segments of *Haynaldia villosa* or *A. speltoides* in the Chinese Spring background were tolerant during the grain-filling stage. Overall a systematic evaluation strategy is needed to assess wild species' potential for identifying traits that could be transferred for better high-temperature tolerance during the high day and night temperatures in field crops.

8.6 FUTURE DIRECTIONS

Proper progression of reproductive stages is responsible for efficient grain yields in field crops. However, this stage is also more sensitive to high-temperature stress during the day and night. Tolerance strategies have till now mainly focused on the limited gene pool in popular cultivars in breeding programs for major crops. As illustrated above, novel donors with higher heat tolerance or escape provide ample evidence for systematically exploring wild species and accessions. Heat or high-temperature stress tolerance could be investigated under its avoidance, escape, and tolerance traits/alleles/genes and underlying phenomenon under field conditions. Future studies on reproductive stage stress in crops must focus on different phenotyping approaches for escape, avoidance, and tolerance traits. It could help to build predictive models with sensitivity timing and threshold temperatures beyond which the reproductive process fails and results in yield loss. New-generation approaches such as lipidomics (for pollen as well as stigma) and metabolomics should be used to complement breeding and plant phenotyping. Sensors for high-throughput phenotyping platforms and drones will vastly improve field-based phenotyping for high-temperature resilient crop breeding programs. Understanding the complexity of this stress at the field level and choosing specific phenotyping programs will help us identify and incorporate traits specific for reproductive stage tolerance during day and night warming in crops.

ACKNOWLEDGMENTS

Pranav Pankaj Sahu acknowledges the grant from the Czech Science Foundation (Grant No. 20-25845Y). Swati Puranik acknowledges the support by the project SustES "Adaptation strategies for sustainable ecosystem services and food security under adverse environmental conditions" (CZ. 02.1.01/0.0/0.0/16_019/0000797).

KEYWORDS

- **reproductive stage**
- **pollen development**
- **heat stress**
- **night-time warming**
- **wild species.**

REFERENCES

1. Hedhly, A. Sensitivity of Flowering Plant Gametophytes to Temperature Fluctuations. *Environ. Exp. Bot.* **2011,** *74*, 9–16.
2. Prasad, P. V. V.; Djanaguiraman, M. Response of Floret Fertility and Individual Grain Weight of Wheat to High Temperature Stress: Sensitive Stages and Thresholds for Temperature and Duration. *Funct. Plant Biol.* **2014,** *41*, 1261–1269.
3. Prasad, P. V.; Bheemanahalli, R.; Jagadish, S. K. Field Crops and the Fear of Heat Stress—Opportunities, Challenges and Future Directions. *Field Crops Res.* **2017,** *200*, 114–121.
4. IPCC. Working Group I Contribution to the IPCC Fifth Assessment Report Climate Change 2013: The Physical Science Basis, Summary for Policymakers, 2013. http://www.climatechange2013.org/images/report/WG1AR5 SPM FINAL.pdf (accessed on 20 June 2016.).
5. Bahuguna, R. N.; Solis, C. A.; Shi, W.; Jagadish, K. S. V. Post-Flowering Night Respiration and Altered Sink Activity Account for High Night Temperature-Induced Grain Yield and Quality Loss in Rice (*Oryza sativa*). *Physiol. Plant.* **2016**. http://dx.doi.org/10.1111/ppl.12485.
6. Garcia, G. A.; Dreccer, M. F.; Miralles, D. J.; Serrago, R. A. High Night Temperatures During Grain Number Determination Reduce Wheat and Barley Grain Yield: A Field Study. *Glob. Change Biol.* **2015,** *21*, 4153–4164.

7. Garcia, G. A.; Serrago, R. A.; Dreccer, M. F.; Miralles, D. J. Post-Anthesis Warm Nights Reduce Grain Weight in Field-Grown Wheat and Barley. *Field Crops Res.* **2016,** *195*, 50–59.
8. Lyman, N. B.; Jagadish, K. S. V.; Nalley, L. L.; Dixon, B. L.; Siebenmorgen, T. Neglecting Rice Milling Yield and Quality Underestimates Economic Losses from High-Temperature Stress. *PLoS One* **2013,** *8* (8), e72157.
9. Prasad, P. V. V.; Djanaguiraman, M. High Night Temperature Decreases Leaf Photosynthesis and Pollen Function in Grain Sorghum. *Funct. Plant Biol.* **2011,** 38, 993–1003.
10. Narayanan, S.; Prasad, P. V. V.; Fritz, A. K.; Boyle, D. L.; Gill, B. S. Impact of High Night Time and High Daytime Temperature Stress on Winter Wheat. *J. Agron. Crop Sci.* **2015,** *201*, 206–218.
11. Shi, W.; Muthurajan, R.; Rahman, H.; Selvam, J.; Peng, S.; Zou, Y.; Jagadish, K. S. V. Source-Sink Dynamics and Proteomic Reprogramming Under Elevated Night Temperature and Their Impact on Rice Yield and Grain Quality. *New Phytol.* **2013,** *197*, 825–837.
12. Sunoj, J. V. S.; Shroyer, K. J.; Jagadish, S. V. K.; Vara Prasad, P. V. Diurnal Temperature Amplitude Alters Physiological and Growth Response of Maize (*Zea mays* L.) During Vegetative Stage. *Environ. Exp. Bot.* **2016,** *130*, 113–121.
13. Welch, J. R.; Vincent, J. R.; Auffhammer, M.; Moya, P. F.; Dobermann, A.; Dawe, D. Rice Yields in Tropical/Subtropical Asia Exhibit Large But Opposing Sensitivities to Minimum and Maximum Temperatures. *Proc. Natl. Acad. Sci. USA* **2010,** *107*, 14562–14567.
14. Prasad, P. V. V.; Djanaguiraman, M.; Perumal, R.; Ciampitti, I. A. Impact Of High Temperature Stress on Floret Fertility and Individual Grain Weight of Grain Sorghum: Sensitive Stages and Thresholds for Temperature and Duration. *Front. Plant Sci.* **2015,** *6*, 820.
15. Shi, W.; Ishimaru, T.; Gannaban, R. B.; Oane, W.; Jagadish, S. V. K.; Popular rice (*Oryza sativa* L.) Cultivars Show Contrasting Responses to Heat Stress at Gametogenesis and Anthesis. *Crop Sci.* **2015,** *55*, 589–596.
16. Singh, V.; Nguyen, C. T.; van-Oosterom, E. J.; Chapman, S. C.; Jordan, D. R.; Hammer, G. L. Sorghum Genotypes Differ in High Temperature Responses for Seed Set. *Field Crops Res.* **2015,** *171*, 32–40.
17. Djanaguiraman, M.; Prasad, P. V. V.; Murugan, M.; Reddy, U. K. Physiological Differences Among Sorghum (*Sorghum bicolor* L. Moench) Genotypes Under High Temperature Stress. *Environ. Exp. Bot.* **2014,** *100*, 43–54.
18. González-Schain, N.; Dreni, L.; Lawas, L. M. F.; Galbiati, M.; Colombo, L.; Heuer, S.; Jagadish, S. V. K.; Kater, M. M. Genome-Wide Transcriptome Analysis During Anthesis Reveals New Insights in the Molecular Basis of Heat Stress Responses in Tolerant and Sensitive Rice Varieties. *Plant Cell Physiol.* **2015,** *57*, 57–68.
19. Li, X.; Lawas, L. M. F.; Malo, R.; Glaubitz, U.; Erban, A.; Mauleon, R.; Heuer, S.; Zuther, E.; Kopka, J.; Hincha, D. K.; Jagadish, S. V. K. Metabolic and Transcriptomic Signatures of Rice Floral Organs Reveal Sugar Starvation as a Factor in Reproductive Failure Under Heat and Drought Stress. *Plant Cell Environ.* **2015,** *38*, 2171–2192.
20. Polowick, P. L.; Sawhney, V. K. High-Temperature Induced Male and Female Sterility in Canola (*Brassica napus* L.). *Ann. Bot.* **1988,** *62*, 83–86.

21. Saini, H. S.; Sedgley, M.; Aspinall, D. Effect of Heat Stress During Floral Development on Pollen Tube Growth and Ovary Anatomy in Wheat *(Triticum aestivum* L.). *Aust. J. Plant Physiol.* **1983,** *10*, 137–144.
22. Jagadish, S. V. K.; Muthurajan, R.; Oane, R.; Wheeler, T. R.; Heuer, S.; Bennett, J.; Craufurd, P. Q. Physiological and Proteomic Approaches to Dissect Reproductive Stage Heat Tolerance in Rice (*Oryza sativa* L.). *J. Exp. Bot.* **2010,** *61*, 143–156.
23. Kumar, S.; Thakur, P.; Kaushal, N.; Malik, J. A.; Gaur, P.; Nayyar, H. Effect of Varying High Temperatures During Reproductive Growth on Reproductive Function: Oxidative Stress and Seed Yield in Chickpea Genotypes Differing in Heat Sensitivity. *Arch. Agron. Soil Sci.* **2013,** *59*, 823–843.
24. Snider, J. L.; Oosterhuis, D. M.; Skulman, B. W.; Kawakami, E. M. Heat Stress Induced Limitations to Reproductive Success in *Gossypium hirsutum*. *Physiol. Plant* **2009** *137*, 125–138.
25. Snider, J. L.; Oosterhuis, D. M.; Loka, D. A.; Kawakami, E. M. High Temperature Limits In Vivo Pollen Tube Growth Rates by Altering Diurnal Carbohydrate Balance in Field-Grown *Gossypium hirsutum* Pistils. *J. Plant Physiol.* **2011,** *168*, 1168–1175.
26. Kaushal, N.; Bhandari, K.; Siddique, K. H.; Nayyar, H. Food Crops Face Rising Temperatures: An Overview of Responses, Adaptive Mechanisms, and Approaches to Improve Heat Tolerance. *Cogent Food Agric.* **2016,** *2*, p.1134380.
27. Hirabayashi, H.; Sasaki, K.; Kambe, T.; Gannaban, R. B.; Miras, M. A.; Mendioro, M. S.; Simon, E. V.; Lumanglas, P. D.; Fujita, D.; Takemoto-Kuno, Y.; Takeuchi, Y.; Kaji, R.; Kondo, M.; Kobayashi, N.; Ogawa, T.; Ando, I.; Jagadish, K. S. V.; Ishimaru, T. qEMF3, a Novel QTL for Early-Morning Flowering Trait from Wild Rice, *Oryza officinalis*, to Mitigate Heat Stress Damage at Flowering in Rice (*Oryza sativa* L.). *J. Exp. Bot.* **2014,** *66*, 1227–1236.
28. Julia, C.; Dingkuhn, M. Predicting Temperature Induced Sterility of Rice Spikelets Requires Simulation of Crop-Generated Microclimate. *Eur. J. Agron.* **2013,** *49*, 50–60.
29. Pradhan, G. P.; Prasad, P. V. V.; Fritz, A. K.; Kirkham, M. B.; Gill, B. S. High Temperature Tolerance in Aegilops Species and Its Potential Transfer to Wheat. *Crop Sci.* **2012,** *52*, 292–304.
30. Pradhan, G. P.; Prasad, P. V. V. Evaluation of Wheat Chromosomal Translocation Lines for High Temperature Stress Tolerance During Grain Filling Stage. *PLoS One* **2015,** *10*, e0116620.
31. Bahuguna, R. N.; Jha, J.; Madan, P.; Shah, D.; Lawas, M. L.; Khetarpal, S.; Jagadish, S. V. K. Physiological and Biochemical Characterization of NERICA-L-44: A Novel Source of Heat Tolerance at the Vegetative and Reproductive Stages in Rice. *Physiol. Plant.* **2015,** *154*, 543–559.
32. Prasad, P. V. V.; Craufurd, P. Q.; Kakani, V. G.; Wheeler, T. R.; Boote, K. J.; Influence of High Temperature During Pre- and Post-Anthesis Stages of Floral Development on Fruit-Set and Pollen Germination in Peanut. *Aust. J. Plant. Physiol.* **2001,** *28*, 233–240.
33. Prasad, P. V. V.; Djanaguiraman, M.; Perumal, R.; Ciampitti, I. A. Impact of High Temperature Stress on Floret Fertility and Individual Grain Weight of Grain Sorghum: Sensitive Stages and Thresholds for Temperature and Duration. *Front. Plant Sci.* **2015,** *6*, 820.

34. Yoshida, S.; Satake, T.; Mackill, D. High Temperature Stress in Rice. In: IRRI Research Paper Series 67. *Int. Rice Research Inst.*, 1981, Manila.
35. Jagadish, S. V. K.; Craufurd, P. Q.; Wheeler, T. R. Phenotyping Rice Mapping Population Parents for Heat Tolerance During Anthesis. *Crop Sci.* **2008,** *48*, 1140–1146.
36. Prasad, P. V. V.; Craufurd, P. Q.; Summerfield, R. J.; Wheeler, T. R. Effects of Short Episodes of Heat Stress on Flower Production and Fruit-Set of Groundnut (*Arachis hypogaea* L.). *J. Exp. Bot.* **2000,** *51*, 777–784.
37. Dubey, R.; Pathak, H.; Chakrabarti, B.; Singh, S.; Gupta, D. K.; Harit, R. C.; Impact of Terminal Heat Stress on Wheat Yield in India and Options for Adaptation. *Agric. Syst.* **2020,** *181*, 102826.
38. Julia, C.; Dingkuhn, M. Variation in Time of Day of Anthesis in Rice in Different Climatic Environments. *Euro. J. Agron.* **2012,** *43*, 166–174.
39. Weerakoon, W. M. W.; Maruyama, A.; Ohba, K. Impact of Humidity on Temperature-Induced Grain Sterility in Rice (*Oryza sativa* L.). *J. Agron. Crop Sci.* **2008,** *194*, 135–140.
40. Ye, C.; Tenorio, F. A.; Redona, E. D.; Morales-Cortezano, P. S.; Cabrega, G. A.; Jagadish, K. S. V.; Gregorio, G. B. Fine-Mapping and Validating qHTSF4.1 to Increase Spikelet Fertility Under Heat Stress at Flowering in Rice. *Theor. Appl. Genet.* **2015,** *128*, 1507–1517.
41. Suzuki, K.; Tsukaguchi, T.; Takeda, H.; Egawa, Y. Decrease of Pollen Sustainability of Green Bean at High Temperatures and Relationship to Heat Tolerance. *J. Am. Soc. Hortic. Sci.* **2001,** *126*, 571–574.
42. Salem, M. A.; Kakani, V. G.; Koti, S.; Reddy, K. R. Pollen-Based Screening of Soybean Genotypes for High Temperatures. *Crop Sci.* **2007,** *47*, 219–231.
43. Sakata, T.; Oshino, T.; Miura, S.; Tomabechi, M.; Tsunaga, Y.; Higashitani, N.; Miyazawa, Y.; Takahashi, H.; Watanabe, M.; Higashitani, A. Auxins Reverse Plant Male Sterility Caused by High Temperatures. *Proc. Natl. Acad. Sci. USA* **2010,** *107*, 8569–8574.
44. Sage, T. L.; Bagha, S.; Lundsgaard-Nielsen, V.; Branch, H. A.; Sultmanis, S.; Sage, R. F. The Effect of High Temperature Stress on Male and Female Reproduction in Plants. *Field Crops Res.* **2015,** *182*, 30–42.
45. Mamun, E. A.; Cantrill, L. C.; Overall, R. L.; Sutton, B. G. Cellular Organisation in Meiotic and Early Post-Meiotic Rice Anthers. *Cell Biol. Int.* **2005,** *29*, 903–913.
46. Dolferus, R.; Ji, X.; Richards, R. A. Abiotic Stress and Control of Grain Number in Cereals. *Plant Sci.* **2011,** *181*, 331–341.
47. Kumar, R. R.; Goswami, S.; Gadpayle, K. A.; Singh, K.; Sharma, S. K.; Singh, G. P.; Pathak, H.; Rai, R. D. Ascorbic Acid at Pre-Anthesis Modulate the Thermotolerance Level of Wheat (*Triticum aestivum*) Pollen Under Heat Stress. *J. Plant Biochem. Biotechnol.* **2014,** *23*, 293–306.
48. Awasthi, R.; Bhandari, K.; Nayyar, H. Temperature Stress and Redox Homeostasis in Agricultural Crops. *Front. Environ. Sci.* **2015,** *3*, 11.
49. Parrotta, L.; Faleri, C.; Cresti, M.; Cai, G. Heat Stress Affects the Cytoskeleton and the Delivery of Sucrose Synthase in Tobacco Pollen Tubes. *Planta* **2016,** *243*, 43–63.
50. Jain, M.; Prasad, P. V. V.; Boote, K. J.; Allen, L. H.; Chourey, P. S. Effects of Season-Long High Temperature Growth Conditions on Sugar-to-Starch Metabolism in

Developing Microspores of Grain Sorghum (*Sorghum bicolor* L. Moench). *Planta* **2007,** *227*, 67–79.
51. Jain, M.; Chourey, P. S.; Boote, K. J.; Allen, L. H.; Short-Term High Temperature Growth Conditions During Vegetative-to-Reproductive Phase Transition Irreversibly Compromise Cell Wall Invertase-Mediated Sucrose Catalysis and Microspore Meiosis in Grain Sorghum (*Sorghum bicolor*). *J. Plant Physiol.* **2010,** *167*, 578–582.
52. Parish, R. W.; Phan, H. A.; Iacuone, S.; Li, S. F. Tapetal Development and Abiotic Stress: A Center of Vulnerability. *Funct. Plant Biol.* **2012,** *39*, 553–559.
53. Allen, A. M.; Thorogood, C. J.; Hegarty, M. J.; Lexer, C.; Hiscock, S. J.; Pollen–Pistil Interactions and Self-Incompatibility in the Asteraceae: New Insights from Studies of *Senecio squalidus* (*Oxford ragwort*). *Ann. Bot.* **2011,** *108*, 687–698.
54. Gupta, S. K.; Rao, K. N.; Singh, P.; Ameta, V. L.; Gupta, S. K.; Jayalekha, A. K.; Mahala, R. S.; Pareek, S.; Swamin, M. L.; Verma, Y. S. Seed Set Variability Under High Temperatures During Flowering Period in Pearl Millet (*Pennisetum glaucum* L(R.) Br.). *Field Crops Res.* **2015,** *171*, 41–53.
55. Mohammed, A. R.; Tarpley, L. Differential Response of Two Important Southern US Rice (*Oryza sativa* L.) Cultivars to High Night Temperature. *Aust. J. Crop Sci.* **2014,** *8*, 191–199.
56. Coast, O.; Ellis, R. H.; Murdoch, A. J.; Quinones, C.; Jagadish, K. S. V.; High Night Temperature Induces Contrasting Responses for Spikelet Fertility Spikelet Tissue Temperature, Flowering Characteristics and Grain Quality in Rice. *Funct. Plant Biol.* **2014,** *42*, 149–161.
57. Shi, W.; Xiao, G.; Struik, P. C.; Jagadish, K. S. V.; Yin, X. Quantifying Source-Sink Relationships of Rice Under High Night-Time Temperature Combined with Two Nitrogen Levels. *Field Crops Res.* **2016.** http://dx.doi.org/10.1016/j.fcr.2016.05.013.
58. Ishimaru, T.; Hirabayashi, H.; Ida, M.; Takai, T.; San-Oh, Y. A.; Yoshinaga, S.; Ando, I.; Ogawa, T.; Motohiko, K. A Genetic Resource for Early-Morning Flowering Trait of Wild Rice *Oryza officinalis* to Mitigate High Temperature-Induced Spikelet Sterility at Anthesis. *Ann. Bot.* **2010,** *106*, 515–520.
59. Satake, T.; Yoshida, S. High Temperature Induced Sterility in Indica Rice at Flowering [in Japanese, English Summary]. *Jpn. J. Crop Sci.* **1978,** *47*, 6–17.
60. Jagadish, S. V. K.; Cairns, J. E.; Kumar, A.; Somayanda, I. M.; Craufurd, P. Q. Does Susceptibility to Heat Stress Confound Screening for Drought Tolerance? *Funct. Plant Biol.* **2011,** *38*, 261–269.

CHAPTER 9

Effect of Night-time Warming on Yield and Quality Attributes

RAJEEV KUMAR[1], DATA RAM SAINI[2], DASARI SREEKANTH[3], and T. K. BEHERA[1]

[1]*ICAR-Indian Institute of Vegetable Research, Varanasi, India*

[2]*Institute of Agricultural Sciences, Banaras Hindu University, Varanasi, India*

[3]*ICAR-Directorate of Weed Research, Jabalpur, India*

ABSTRACT

The asymmetric rise in temperature, with a faster increase in daily minimum temperature than the daily maximum temperature, which creates "high night temperature" (HNT) conditions, has emerged as one of the most important limiting factors for enhancing crop production across the globe. The decrease in the yield of crops under HNT is governed by complex physiological mechanisms operating at different tissues, organs, and organizational levels. HNT leads to increased night-time respiration and triggers the deactivation of several key chloroplastic stroma enzymes that further downregulate photosynthesis in the following day and trigger leaf senescence, which ultimately limits the carbon availability to the growing parts. Apart from this, HNT also causes increased accumulation of reactive oxygen species (ROS), leading to increased membrane damage, ultimately leading to lower pollen viability, lower seed set, and lower yield. Moreover, HNT also affects the quality attributes of the crop end products as it has been widely reported in the case of wheat and rice

Impact of High Night Temperature on Plant Biology: Toward Sustainable Plant Adaptation to Climate Change. Rajesh Kumar Singhal, Raju Bheemanahalli, Saurabh Pandey, and Pratibha M. D. (Eds.)

that HNT causes alteration in grain composition, particularly in terms of starch and protein content, which determine the grain quality. Hence, to ensure global food security, there is an urgent need for the development of climate-resilient cultivars, which have HNT tolerance ability.

9.1 INTRODUCTION

In recent times, climate change has emerged as one of the most important limiting factors for enhancing crop production across the globe. Because of this, there has been an increased incidence of abiotic and biotic stresses that negatively affect the yield of major growing crops.[1,2] During the last century, the average global surface temperature has increased by 0.85°C, and it has been predicted that it will increase up to 3.7°C by 2100.[2] Furthermore, the major concern is that there has been an asymmetric rise in temperature, with a faster increase in the daily minimum temperature than the daily maximum temperature, which creates "high night temperature" (HNT) conditions.[3] Central and South Asia, or North America, are the major crop-growing regions where increased night-time temperatures have appeared as a severe problem.[4–8] This increased HNT due to global warming leads to decreased crop yield.[3] Several studies have been conducted to investigate the impact of HNT on the growth and yield of cereals, particularly rice and wheat.[9–13] This chapter emphasizes on the effect of night-time warming temperatures on yield and yield attributes with respect to cereals.

9.2 HNT IMPACT ON YIELD AND TRAITS ASSOCIATED WITH YIELD IN CEREALS

9.2.1 TOTAL GRAIN YIELD

High night temperature (3.9°C increase over ambient) incidence during anthesis and grain-filling stages in wheat and barley leads to 7 and 4% yield reductions, respectively, per °C increase in night temperature.[9,10] At the anthesis stage, the reduction in yield is mainly due to reduced seed set and grain number spike,$^{-1}$ whereas at the grain-filling stage, it is mainly due to a significant decrease in kernel weight due to the shortened grain-filling period[14] in wheat and barley. It has been reported that there is an estimated 6.2–10% loss in paddy yield for every 1°C rise in the whole-season minimum night temperature.[3,15]

9.2.2 GRAIN NUMBER

Grain number is considered a major component of yield in cereals, particularly rice and wheat.[16] High-temperature stress enhances spike development and decreases spikelets' number and grains per spike.[17] The reduced grain number per spike in wheat is mainly due to pollen sterility. In the case of wheat, temperatures above 31°C just before anthesis cause pollen sterility in wheat. A 4% yield reduction as a grain number per unit area for each °C increase (from 15 to 22°C) in mean temperature has been reported before anthesis in wheat crops.[18]

9.2.3 INDIVIDUAL GRAIN WEIGHT

The duration between anthesis and physiological maturity decreases with increasing temperature, which negatively impacts individual grain weight. It is also reported in wheat that high temperatures (32/27°C) from anthesis to maturity cause a 20% reduction in the average grain weight. Under high-temperature stress conditions, the individual grain weight reduction is due to the decrease in the size of endosperm cells, and reduced starch deposition in grains has also been reported in some crops.

9.2.4 GRAIN-FILLING DURATION

The grain-filling stage is the most sensitive to high-temperature stress in most growing crops. The development of grain is accompanied by the synthesis and accumulation of storage substances such as lipids, starch, and storage proteins, which are determining factors for rice/wheat grain quality. High-temperature stress mainly promotes the rate of grain-filling and declines its active duration, resulting in the reduction of photosynthetic substance accumulation in rice grains.[19,20] High-temperature stress is also responsible for the reduction of individual grain weight due to shortening the grain-filling duration. The grain-filling duration decreases by 2–8 days for every 1°C increase above 15–20°C in wheat.[21] The grain-filling period shortens by 3 and 7 days at night temperatures of 20 and 23°C, respectively, compared with a night temperature of 14°C in wheat crops.[22]

9.2.5 GRAIN-FILLING RATE

The decreased grain-filling duration under optimum temperature is generally compensated by an increased grain-filling rate, but under high-temperature stress, this compensation does not occur, resulting in a reduction of individual grain weight. The major reasons for decreased grain-filling rates at high temperatures are decreased leaf and spike photosynthesis and decreased stem reserve mobilization.

9.3 HNT IMPACT ON GRAIN PHYSICAL CHARACTERISTICS, COMPOSITION, AND BIOCHEMICAL CHANGES

9.3.1 PHYSICAL CHARACTERISTICS OF WHEAT AND RICE GRAINS

High night temperatures alter carbon balance or cause a reduction in assimilate supply, favoring carbon loss through night-time respiration to the later flowering florets, which resulted in the formation of shrunken grains (reduced length and width) both in rice and wheat, due to a decrease in endosperm cell area or a decrease in the accumulation of starch in the endosperm.[13,23,24]

9.3.2 STARCH, PROTEIN, AND LIPID COMPOSITION CHANGES IN DEVELOPING WHEAT AND RICE GRAINS AND END-USE QUALITY

High-temperature stress causes an alteration in grain composition, particularly in terms of starch and protein content, which determines the grain quality in wheat and rice.[25,26] It has been reported that HNT (>35°C) during the grain-filling period leads to reduced starch accumulation and favors protein and lipid accumulation in developing wheat seeds.[13] Due to the shortened grain-filling duration under heat stress, there are predominantly larger type A starch granules in wheat seeds, which are accumulated early in grain development, whereas smaller type B starch granules are predominant in control conditions and are accumulated in the mid-stages of grain development. The presence of larger type A starch granules exhibits pitting in wheat seeds.[27,28] Moreover, the amylose-to-amylopectin ratio also increased under heat stress in wheat, which affects starch gelation properties, gelatinization, and pasting, resulting in lower dough elasticity.[28,29] As the protein percentage increases in wheat flour due to heat stress, there is an increased loaf volume and a reduction in mixing tolerance, reflecting lower water absorption and a shorter stabilization time of the dough.[30] HNT also increases the ratio of α-gliadin and glutenin content in wheat grains, which negatively affects the dough property and noodle swelling power of the wheat flour.[31] Moreover, lipid accumulation also increased in wheat grains under HNT,[13] which could promote the oxidation of free fatty acids and decrease the bread-baking quality of byproducts with undesirable aroma and taste.[32]

Similarly, in the case of rice, HNT during grain-filling causes a reduction in starch accumulation and increases the accumulation of storage proteins, with a significantly increased glutelin composition and a decrease in prolamine in rice grains. The increased protein content in rice grain decreases its eating quality and makes it more susceptible to spoilage during storage.[33] The protease activity is enhanced under high temperatures, and it further enhances the protein transformation into soluble nitrogen compounds, which increases the total amino acids in rice grains. The reduction in starch accumulation (low amylose content) in rice grains under HNT stress results in increased chalkiness and reduced milling quality in mature grains.[34,35] Chalk formation in rice grains is a result of loosely packed starch granules creating air spaces between amyloplasts, which increases the broken grain percentage upon milling and significantly lowers the market value of the grain.[15,36] Thus, the grain composition alteration in rice grain under HNT led to a reduction of viscosity and increased gelatinization, consequently lowering consumer preference.[37]

9.3.3 ENZYMATIC AND METABOLIC CHANGES IN DEVELOPING WHEAT AND RICE GRAINS

Expression profiling of enzymes involved in starch metabolism in developing wheat seeds revealed lower transcript levels of ADP-glucose pyrophosphorylase and higher transcript levels of starch degradation enzymes such as isoamylase III, alpha-amylase, and beta-amylase, which serve as bottlenecks for starch accumulation.[13] Furthermore, the characterization of enzymes involved in starch metabolism in developing rice grains under HNT revealed that there is a lower activity of starch-synthesizing enzymes such as cell-wall invertase and sucrose synthase, whereas the higher activity of starch-degrading enzymes, α-amylase, results in chalky grains.[35,38] The comparative study of the enzymes involved in starch metabolism in rice and wheat under HNT revealed that most of the enzymes showed a similar trend except sucrose synthase, which increased in wheat but decreased in rice.

9.4 EFFECT OF HNT ON OTHER CROPS

Potato tuber induction and development are sensitive to high temperatures that result in reduced yield in potatoes.[39] In potatoes, the highest harvesting index (HI) of 0.8 and the lowest zero are recorded at night temperatures of 15 and 28°C, respectively, in Northern Indian Uttar Pradesh, Punjab, Haryana, Bihar, and the Northern Hills. A moderate HI of 0.4–0.6 is recorded at night temperatures of 20°C in Central Indian states like Chattisgarh, Gujarat, and some parts of Maharashtra and West Bengal, indicating high night temperatures limiting the partitioning of photosynthesis products to the tubers. A low HI of 0.2 is recorded at more than 20°C night temperatures in South India.[40,41]

The anthesis and pod development are the most sensitive stages to night temperature in snap beans (*Phaseolus vulgaris* L.). Under 32/27°C day/night temperatures, the highest reduction in pod set is recorded due to enhanced abscission of flower buds, flowers, and young pods. Pods larger than 3 cm did not abscise but usually aborted and shriveled under high night temperatures in snap beans (*Phaseolus vulgaris* L.).[42]

Cowpeas are more sensitive to high night-time temperatures than to daytime temperatures[43] because they can tolerate temperatures of 30°C during the day but not at night. In the cowpea, high night temperatures are especially damaging to floral development and pod set.[44] High night temperatures also decrease specific leaf weight without affecting leaf area due to the depletion of reserve starch and other carbohydrates during the dark period by respiratory losses. The lowest photosynthetic rates are also recorded at 30/25°C compared to 30/15°C in the Wando pea and cowpea.[45] The increase in night temperature over 20.3°C during the reproductive stage of soybeans leads to a reduction in yield of approximately 4.6% per °C, and this decrease in yield can be further increased to 5.1% per °C under extreme HNT from 22 to 28°C.[46]

9.5 REASONS FOR REDUCED YIELD UNDER HIGH NIGHT TEMPERATURE

The decrease in the yield of crops under HNT is governed by complex physiological mechanisms operating at different tissues, organs, and organizational levels (Fig. 9.1).

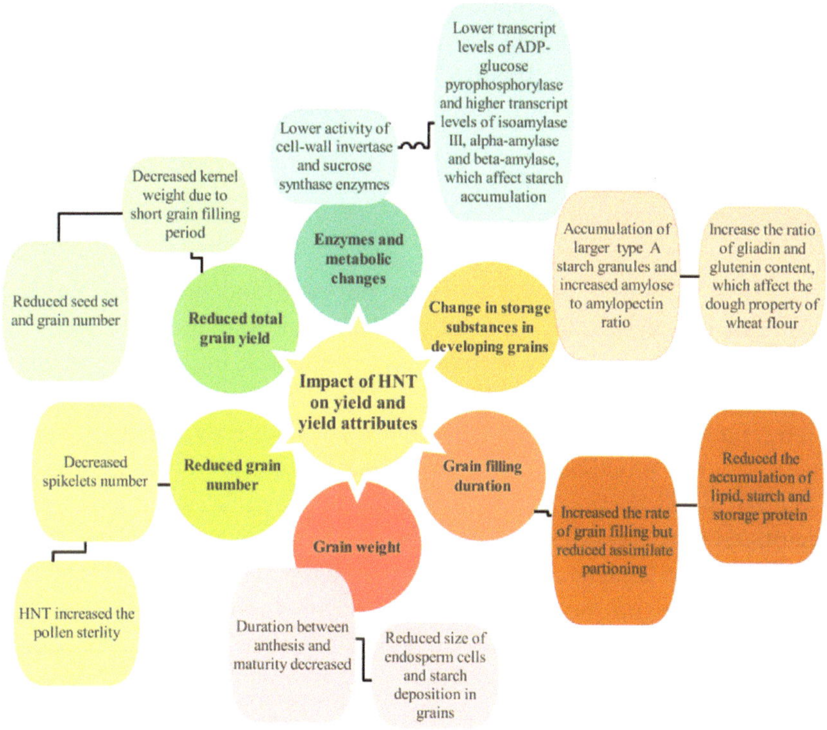

FIGURE 9.1 Effect of high night temperature stress on vegetative and reproductive phases of crops and their impact on yield.

1. HNT causes increased night-time respiration,[24,47] which leads to a reduction of photoassimilates available for plant growth and yield.[10,12] The metabolite profiling studies in wheat and rice under HNT revealed an increase in tricarboxylic acid (TCA) cycle intermediates in leaves, supporting increased respiration in the photosynthesizing tissue.[48]

2. HNT leads to the deactivation of several key chloroplastic stroma enzymes, which further downregulate photosynthesis in the following day and trigger leaf senescence, which ultimately limits the carbon availability to the growing parts.[49,50]
3. HNT causes increased accumulation of reactive oxygen species (ROS), leading to increased membrane damage, ultimately leading to lower pollen viability[51,52] and a lower seed set.
4. HNT during the reproductive stage and increased respiration in reproductive tissues decrease the amount of carbon available for starch production and accumulation, particularly during seed filling, consequently leading to lower seed mass.[10]

9.6 CONCLUSIONS

HNT has a more significant impact on yield and quality. HNT is an alarming concern that needs to be considered since it directly affects yield and quality in a situation where we have to generate high-yield and quality foods to sort out global hunger and meet the demands of global food hunger and ensure food security. HNT stress not only affects the plant's physiological processes that happen during the night but also influences some of the processes occurring during the daytime, such as the photosynthesis rate. Hence, to develop crop varieties that can tolerate HNT, it is essential to understand the direct impact of HNT on night respiration (carbon loss) and the indirect impact on photosynthesis (carbon gain), which can maintain a favorable carbon balance and minimize the negative impact of HNT during grain-filling.

KEYWORDS

- **high night temperature**
- **reactive oxygen species**
- **source–sink relationship**
- **yield attributes**

REFERENCES

1. FAO. FAOSTAT Database. Rome, 2014.
2. IPCC. *AR5 Climate Change 2014: Impacts, Adaptation, and Vulnerability*; Cambridge University Press: Cambridge, 2014.
3. Peng, S.; Huang, J.; Sheehy, J. E.; Laza, R. C.; Visperas, R. M.; Zhong, X.; Cassman, K. G. Rice Yields Decline with Higher Night Temperature from Global Warming. *Proc. Natl. Acad. Sci.* **2004,** *101* (27), 9971–9975.
4. Du, Z.; Zhao, J.; Pan, H.; Wu, Z.; Zhang, H. Responses of Vegetation Activity to the Daytime and Night-time Warming in Northwest China. *Environ. Monitor. Assess.* **2019,** *191* (12), 1–10.
5. Ma, L.; Xia, H.; Meng, Q. Spatiotemporal Variability of Asymmetric Daytime and Night-Time Warming and Its Effects on Vegetation in the Yellow River Basin from 1982 to 2015. *Sensors* **2019,** *19* (8), 1832.
6. Peng, S.; Piao, S.; Ciais, P.; Myneni, R. B.; Chen, A.; Chevallier, F.; Zeng, H. Asymmetric Effects of Daytime and Night-Time Warming on Northern Hemisphere Vegetation. *Nat.* **2013,** *501* (7465), 88–92.
7. Rao, B. B.; Chowdary, P. S.; Sandeep, V. M.; Rao, V. U. M.; Venkateswarlu, B.; Rising Minimum Temperature Trends Over India in Recent Decades: Implications for Agricultural Production. *Glob. Planet. Chang* **2014,** *117*, 1–8.
8. Zhao, X.; Fitzgerald, M. Climate Change: Implications for the Yield of Edible Rice. *PloS One* **2013,** *8* (6), e66218.
9. García, G. A.; Dreccer, M. F.; Miralles, D. J.; Serrago, R. A. High Night Temperatures During Grain Number Determination Reduce Wheat and Barley Grain Yield: A Field Study. *Glob. Chang. Biol.* **2015,** *21* (11), 4153–4164.
10. García, G. A.; Serrago, R. A.; Dreccer, M. F.; Miralles, D. J. Post-Anthesis Warm Nights Reduce Grain Weight in Field-Grown Wheat and Barley. *Field Crops Res.* **2016,** *195*, 50–59.
11. Hein, N. T.; Wagner, D.; Bheemanahalli, R.; Šebela, D.; Bustamante, C.; Chiluwal, A.; Neilsen, M. L.; Jagadish, S. V. Integrating Field-Based Heat Tents and Cyber-Physical System Technology to Phenotype High Night-Time Temperature Impact on Winter Wheat. *Plant Met.* **2019,** *15* (1), 1–15.
12. Impa, S. M.; Sunoj, V. J.; Krassovskaya, I.; Bheemanahalli, R.; Obata, T.; Jagadish, S. K. Carbon Balance and Source-Sink Metabolic Changes in Winter Wheat Exposed to High Night-Time Temperature. *Plant Cell Environ.* **2019,** *42* (4), 1233–1246.
13. Impa, S. M.; Vennapusa, A. R.; Bheemanahalli, R.; Sabela, D.; Boyle, D.; Walia, H.; Jagadish, S. K. High Night Temperature Induced Changes in Grain Starch Metabolism Alters Starch, Protein, and Lipid Accumulation in Winter Wheat. *Plant Cell Environ.* **2020,** *43* (2), 431–447.
14. Narayanan, S.; Prasad, P. V. V.; Fritz, A. K.; Boyle, D. L.; Gill, B. S. Impact of High Night-Time and High Daytime Temperature Stress on Winter Wheat. *J. Agron. Crop Sci.* **2015,** *201* (3), 206–218.
15. Lyman, N. B.; Jagadish, K. S.; Nalley, L. L.; Dixon, B. L. Siebenmorgen, T.; Neglecting Rice Milling Yield and Quality Underestimates Economic Losses from High-Temperature Stress. *PloS One* **2013,** *8* (8), e72157.

16. Evans, L. T. The Influence of Irradiance Before and After Anthesis on Grain Yield and Its Components in Microcrops of Wheat Grown in a Constant Day Length and Temperature Regime. *Field Crops Res.* **1978,** *1*, 5–19.
17. Farooq, M.; Bramley, H.; Palta, J. A.; Siddique, K. H. Heat Stress in Wheat During Reproductive and Grain-Filling Phases. *Crit. Rev. Plant Sci.* **2011,** *30* (6), 491–507.
18. Wheeler, T. R.; Batts, G. R.; Ellis, R. H.; Hadley, P.; Morison, J. I. L. Growth and Yield of Winter Wheat (*Triticum aestivum*) Crops in Response to CO_2 and Temperature. *J. Agric. Sci.* **1996,** *127* (1), 37–48.
19. Kim, J.; Shon, J.; Lee, C. K.; Yang, W.; Yoon, Y.; Yang, W. H.; Lee, B. W. Relationship Between Grain Filling Duration and Leaf Senescence of Temperate Rice Under High Temperature. *Field Crops Res.* **2011,** *122* (3), 207–213.
20. Song, W. F.; Zhao, L. J.; Zhang, X. M.; Zhang, Y. M.; Li, J. L.; Zhang, L. L.; Song, Q.J.; Zhao, H.B.; Zhang, Y.B.; Zhang, C.L.; Xiao, Z. M. Effect of Timing of Heat Stress During Grain Filling in Two Wheat Varieties Under Moderate and Very High Temperature. *Ind. J. Genet. Plant Breed.* **2015,** *75* (1), 121–124.
21. Streck, N. A. Climate Change and Agroecosystems: The Effect of Elevated Atmospheric CO_2 and Temperature on Crop Growth, Development, and Yield. *Ciência Rural* **2005,** *35*, 730–740.
22. Prasad, P. V. V.; Staggenborg, S. A.; Ristic, Z. Impacts of Drought and/or Heat Stress on Physiological, Developmental, Growth, and Yield Processes of Crop Plants. *Response of Crops to Limited Water: Understanding and Modeling Water Stress Effects on Plant Growth Processes* **2008,** *1*, 301–355.
23. Morita, S.; Yonemaru, J. I.; Takanashi, J. I. Grain Growth and Endosperm Cell Size Under High Night Temperatures in Rice (*Oryza sativa* L.). *Ann. Bot.* **2005,** *95* (4), 695–701.
24. Shi, W.; Muthurajan, R.; Rahman, H.; Selvam, J.; Peng, S.; Zou, Y.; Jagadish, K. S.; Source–Sink Dynamics and Proteomic Reprogramming Under Elevated Night Temperature and Their Impact on Rice Yield and Grain Quality. *New Phytol.* **2013,** *197* (3), 825–837.
25. Dou, Z.; Tang, S.; Li, G.; Liu, Z.; Ding, C.; Chen, L.; Ding, Y. Application of Nitrogen Fertilizer at Heading Stage Improves Rice Quality Under Elevated Temperature During Grain-Filling Stage. *Crop Sci.* **2017,** *57* (4), 2183–2192.
26. Tang, S.; Chen, W.; Liu, W.; Zhou, Q.; Zhang, H.; Wang, S.; Ding, Y. Open-Field Warming Regulates the Morphological Structure, Protein Synthesis of Grain and Affects the Appearance Quality of Rice. *J. Cereal Sci.* **2018,** *84*, 20–29.
27. Hurkman, W. J.; Wood, D. F. High Temperature During Grain Fill Alters the Morphology of Protein and Starch Deposits in the Starchy Endosperm Cells of Developing Wheat (*Triticum aestivum* L.) Grain. *J. Agric. Food Chem.* **2011,** *59* (9), 4938–4946.
28. Hurkman, W. J.; McCue, K. F.; Altenbach, S. B.; Korn, A.; Tanaka, C. K.; Kothari, K. M.; Johnson, E. L.; Bechtel, D.B.; Wilson, J. D.; Anderson, O. D.; DuPont, F. M. Effect of Temperature on Expression of Genes Encoding Enzymes for Starch Biosynthesis in Developing Wheat Endosperm. *Plant Sci.* **2003,** *164* (5), 873–881.
29. Nuttall, J. G., O'leary, G. J., Panozzo, J. F., Walker, C. K., Barlow, K. M., Fitzgerald, G. J. Models of Grain Quality in Wheat—A Review. *Field Crops Res.* **2017,** *202*, 136–145.

30. Dupont, F. M.; Hurkman, W. J.; Vensel, W. H.; Tanaka, C.; Kothari, K. M.; Chung, O. K.; Altenbach, S. B. Protein Accumulation and Composition in Wheat Grains: Effects of Mineral Nutrients and High Temperature. *Eur. J. Agron.* **2006,** *25* (2), 96–107.
31. Rangan, P.; Furtado, A.; Henry, R. Transcriptome Profiling of Wheat Genotypes Under Heat Stress During Grain-Filling. *J. Cereal Sci.* **2020,** *91*, 102895.
32. Barnes, P. Cereal Lipids. *Nutri. Food Sci.* **1984**.
33. Cao, Z.; Zhao, Q.; Pan, G.; Wei, K.; Zhou, L.; Cheng, F. Comprehensive Expression of Various Genes Involved in Storage Protein Synthesis in Filling Rice Grain as Affected by High Temperature. *Plant Growth Regul.* **2017,** *81* (3), 477–488.
34. Ambardekar, A. A.; Siebenmorgen, T. J.; Counce, P. A.; Lanning, S. B.; Mauromoustakos, A. Impact of Field-Scale Night-time Air Temperatures During Kernel Development on Rice Milling Quality. *Field Crops Res.* **2011,** *122* (3), 179–185.
35. Hakata, M.; Kuroda, M.; Miyashita, T.; Yamaguchi, T.; Kojima, M.; Sakakibara, H.; Mitsui, T.; Yamakawa, H. Suppression of α-Amylase Genes Improves Quality of Rice Grain Ripened Under High Temperature. *Plant Biotechnol. J.* **2012,** *10* (9), 1110–1117.
36. Sadok, W.; Jagadish, S. K. The Hidden Costs of Night-time Warming on Yields. *Trends Plant Sci.* **2020,** *25* (7), 644–651.
37. Song, X.; Du, Y.; Song, X.; Zhao, Q.; Effect of High Night Temperature During Grain Filling on Amyloplast Development and Grain Quality in Japonica Rice. *Cereal Chem.* **2013,** *90* (2), 114–119.
38. Yamakawa, H.; Hirose, T.; Kuroda, M.; Yamaguchi, T. Comprehensive Expression Profiling of Rice Grain Filling-Related Genes Under High Temperature Using DNA Microarray. *Plant Physiol.* **2007,** *144* (1), 258–277.
39. Reynolds, M. P.; Ewing, E. E. Effects of High Air and Soil Temperature Stress on Growth and Tuberization in *Solanum tuberosum*. *Ann. Bot.* **1989,** *64* (3), 241–247.
40. Pandey, S. K.; Govindakrishnan, P. M.; Rawat, S. Potato Stress Prone Environments and Challenges. Chadha et al., 2009; pp 72–78.
41. Singh, A. K. Climate Change Sensitivity of Indian Horticulture-Role of Technological Interventions, Souvenir of Fourth Indian Horticultural Congress, HSI, New Delhi. HSI, New Delhi, 2010; pp 85–95.
42. Konsens, I.; Ofir, M.; Kigel, J. The Effect of Temperature on the Production and Abscission of Flowers and Pods in Snap Bean (*Phaseolus vulgaris* L.). *Ann. Bot.* **1991,** *67* (5), 391–399.
43. Mutters, R. G.; Hall, A. E. Reproductive Responses of Cowpea to High Temperature During Different Night Periods. *Crop Sci.* **1992,** *32* (1), 202–206.
44. Hall, A. E. Breeding for Heat Tolerance. *Plant Breed. Rev.* **1992,** *10* (2), 129–168.
45. McDonald, S.; Paulsen, G. M. Combined Effects of Drought and High Temperature on Water Relations of Wheat and Sorghum. *Plant Soil* **2001,** *233* (2), 179–187.
46. Lin, T.; Okamoto, Y.; Nagasaki, Y.; Shiraiwa, T. The Influence of High Night Temperature on Yield and Physiological Attributes of Soybean cv. Fukuyutaka. *Plant Prod. Sci.* **2021,** *24* (3), 267–278.
47. Bahuguna, R. N.; Solis, C. A.; Shi, W.; Jagadish, K. S. Post-Flowering Night Respiration and Altered Sink Activity Account for High Night Temperature-Induced

Grain Yield and Quality Loss in Rice (*Oryza sativa* L.). *Physiologia Plantarum* **2017,** *159* (1), 59–73.

48. Glaubitz, U.; Erban, A.; Kopka, J.; Hincha, D. K.; Zuther, E. High Night Temperature Strongly Impacts TCA Cycle, Amino Acid and Polyamine Biosynthetic Pathways in Rice in a Sensitivity-Dependent Manner. *J. Exp. Bot.* **2015,** *66* (20), 6385–6397.
49. Sakuraba, Y.; Lee, S. H.; Kim, Y. S.; Park, O. K.; Hörtensteiner, S.; Paek, N. C.; Delayed Degradation of Chlorophylls and Photosynthetic Proteins in Arabidopsis Autophagy Mutants During Stress-Induced Leaf Yellowing. *J. Exp. Bot.* **2014,** *65* (14), 3915–3925.
50. Tombesi, S.; Cincera, I.; Frioni, T.; Ughini, V.; Gatti, M.; Palliotti, A.; Poni, S.; Relationship Among Night Temperature, Carbohydrate Translocation and Inhibition of Grapevine Leaf Photosynthesis. *Environ. Exp. Bot.* **2019,** *157*, 293–298.
51. Luria, G.; Rutley, N.; Lazar, I.; Harper, J. F.; Miller, G. Direct Analysis of Pollen Fitness by Flow Cytometry: Implications for Pollen Response to Stress. *Plant J.* **2019,** *98* (5), 942–952.
52. Prasad, P. V.; Djanaguiraman, M. High Night Temperature Decreases Leaf Photosynthesis and Pollen Function in Grain Sorghum. *Funct. Plant Biol.* **2011,** *38* (12), 993–1003.

PART IV

PLANT GROWTH REGULATORS AND OMICS UNDER NIGHT WARMING

CHAPTER 10

The Role of Plant Growth Regulators in Combating High Night Temperature Stress

HEMLATA SINGH, TEIKUR MAJAW, and JYOSTNA RANI PRADHAN

Department of Botany, Plant Physiology & Biochemistry, CBS&H, Dr. Rajendra Prasad Central Agricultural University, Pusa, Bihar, India

ABSTRACT

Global climate warming is a major constraint to crop productivity worldwide. Recent studies indicated that an increase in night temperature is faster than the day temperature, and a slight increase in night temperature can have a detrimental effect on different developmental processes, which eventually decreases crop yield. High night temperature (HNT) is considered unfavorable because it affects whole plant growth, starting from seed germination to grain yield, by disrupting normal plant metabolism, *viz.*, physiology, biochemistry, and gene expression. However, plants have evolved various defense strategies to counteract these adversities. One of the protective mechanisms is the participation of several plant growth regulators (PGRs) in heat stress tolerance. To better understand plant thermotolerance strategies, it is crucial to unravel the intricate web of interactions between distinct PGRs. Studies focusing on the role of PGRs in combating heat stress due to high night temperatures are rare. Hence, the emphasis of this chapter is on the function of plant growth regulators and their signaling under high night-time temperatures. Based on our understanding, we provide a picture of how plants perceive and respond to high temperatures at night via plant growth regulators.

Impact of High Night Temperature on Plant Biology: Toward Sustainable Plant Adaptation to Climate Change. Rajesh Kumar Singhal, Raju Bheemanahalli, Saurabh Pandey, and Pratibha M. D. (Eds.)

10.1 INTRODUCTION

A key factor limiting crop growth and development globally is rising ambient temperature. According to assessments of global climate change,[1] the average temperature will rise by 1–3.7°C by the end of this century in comparison to the average temperature between 1850 and 1900. Temperatures at night are anticipated to rise more quickly than during the day because of reduced radiative heat loss brought on by an increase in cloud cover.[2] Increased temperature induces heat stress in plants. The severity of the heat stress is determined by the duration of the high temperature above the optimum temperature for the crop during the day or night time occurring in a particular climate zone. The majority of research on the effect of increased temperatures assumes that increases in day and night temperatures have equal effects on crop development and grain yield, which obscures the differing impacts of high night and day temperatures.[3] In recent days, a number of investigations have been done on the effect of high night temperatures on different crops,[2–8] which indicated that high night temperatures (HNTs) also have a drastic impact on different plant metabolic processes (Fig. 10.1), leading to disrupted growth and development in plants.

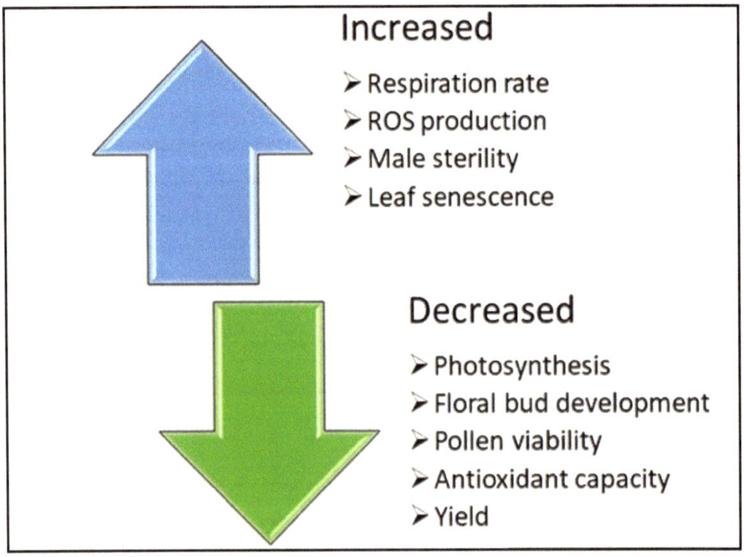

FIGURE 10.1 Impact of high night temperatures on different metabolic functions.

However, plants have developed several mechanisms to deal with such stress. One such mechanism is thermotolerance. Thermotolerance is an inherent characteristic of the plant for its survival.[9] Plants protect themselves by involving both basal and acquired thermotolerance mechanisms. Basal thermotolerance is the innate capacity of plants to endure exposure to temperatures above the optimum temperature necessary for growth, whereas acquired thermotolerance is the capacity of plants to withstand severe heat stress after becoming accustomed to mild stress.[10] Basal thermotolerance is involved in the production and signaling of various PGRs, and it involves the antioxidant system.[11,12] Thermotolerance is acquired by heat shock proteins (HSPs), and various PGRs are involved in the signaling, expression, and regulation of these HSPs.[13,14]

Plant growth regulators (PGRs) are chemical compounds, which are synthesized naturally in plant systems or artificially synthesized as analogues of natural PGRs. PGRs in low concentrations are utilized to promote or hinder plant growth and development, as well as to change specific physiology or metabolic processes (Fig. 10.2). Besides their role in normal metabolism, a diverse group of PGRs like auxin, gibberellic acid, glycine betaine, abscisic acid, ethylene, cytokinins, jasmonic acid, salicylic acid, and brassinosteroids have a significant role in modifying different types of stresses. They protect plants by acting as thermoprotectants, synthesizing stress proteins, enhancing antioxidants, improving photosynthesis efficiency, and performing many other regulatory functions related to metabolic processes.[15,16] Plants have complex mechanisms to respond to stress with the help of PGRs, and their signaling pathways are complexly interconnected. It is important to understand these complex interactions and signaling pathways.

When plants undergo stress, production of these PGRs increases. However, these PGRs become insufficient to protect the plant against the more stressful condition. Many studies showed that exogenous application of PGRs enhanced the plant's tolerance to stress. In this era of climate change, controlling PGR levels through an exogenous application or breeding method may help to minimize heat stress and maintain crop productivity. The role of various PGRs in heat stress due to high temperatures at night is limited and requires further investigation.

10.2 ROLES OF PLANT GROWTH REGULATORS IN RESPONSE

FIGURE 10.2 Role of various plant growth regulators in different physiological and metabolic processes.

TO HEAT STRESS

10.2.1 AUXINS

The phytohormone auxin is acknowledged for mediating morphological changes in plants. Plants undergo substantial morphological and structural changes as a strategy to survive under increasing temperatures, which is known as thermomorphogenesis. Auxin, an important phytohormone type of PGR, is crucial for heat stress-induced thermomorphogenesis, which includes leaf hyponasty and stem (hypocotyl) elongation.[17] Wang et al.[18] reported that HSPs are also involved in morphological and developmental changes mediated by auxin. HSPs regulate temperature growth responses by stimulating auxin signaling rather than modifying auxin biosynthesis.

Auxins are also involved in cell division and elongation. The cotyledons had higher concentrations of transcripts of auxin biosynthetic genes YUCCA8 and YUCCA9 than the hypocotyl, indicating that cotyledons are the primary site of auxin biosynthesis during HT stress, which was confirmed by employing polar transport inhibitors.[19] Oshino et al.[20] suggested that exogenous application of auxin before the HT treatment promoted reproductive growth by reversing male sterility and enhancing anther cell proliferation. Transcription factor Phytochrome-Interacting Factor 4 (PIF4) is necessary for auxin-mediated hypocotyl elongation.[21] Sun et al.[22] reported that auxin levels were not increased in *pif4* mutant plants at high temperatures. The in vitro study showed that PIF4 binds to the promoter region of gene YUCCA8 and upregulates its expression (Fig. 10.3a).

The high night temperature has a different aspect of the expression of various other plant growth hormones, which is different from the effect imposed by the high day temperature. The differential expression of these hormones is reflected in the morphology of the plants as a whole. For instance, in tomato seedlings, the seedlings exposed to high night temperatures show less concentration of gibberellins and auxins, and the seedlings show stunted growth since stem elongation was inhibited, but stem thickness and the number of xylem vessels were normal and not affected when compared to the seedlings exposed to high day temperatures.[23] The mechanism of stem elongation inhibition stems from the inhibition of the biosynthesis of auxins. High night-time temperatures inhibit the expression of YUCCA genes, which are involved in auxin biosynthesis[24] (Fig. 10.3b). The decreased endogenous level of auxins resulted in the inactivation of potassium channels, ultimately decreasing the cytoplasmic potassium concentration, leading to the nonabsorption of water by osmosis and, hence, no cell elongations.[25]

10.2.2 CYTOKININS

Cytokinins (CTKs) are important plant growth regulators that promote cell growth and also mediate heat tolerance. A number of studies are there that suggest that cytokinins (CTKs) are involved in adaptive mechanisms to heat stress.[26–28] Plants subjected to high temperatures show reduced cellular CTK, CTK transportation rate, and high activity of CTK-degrading

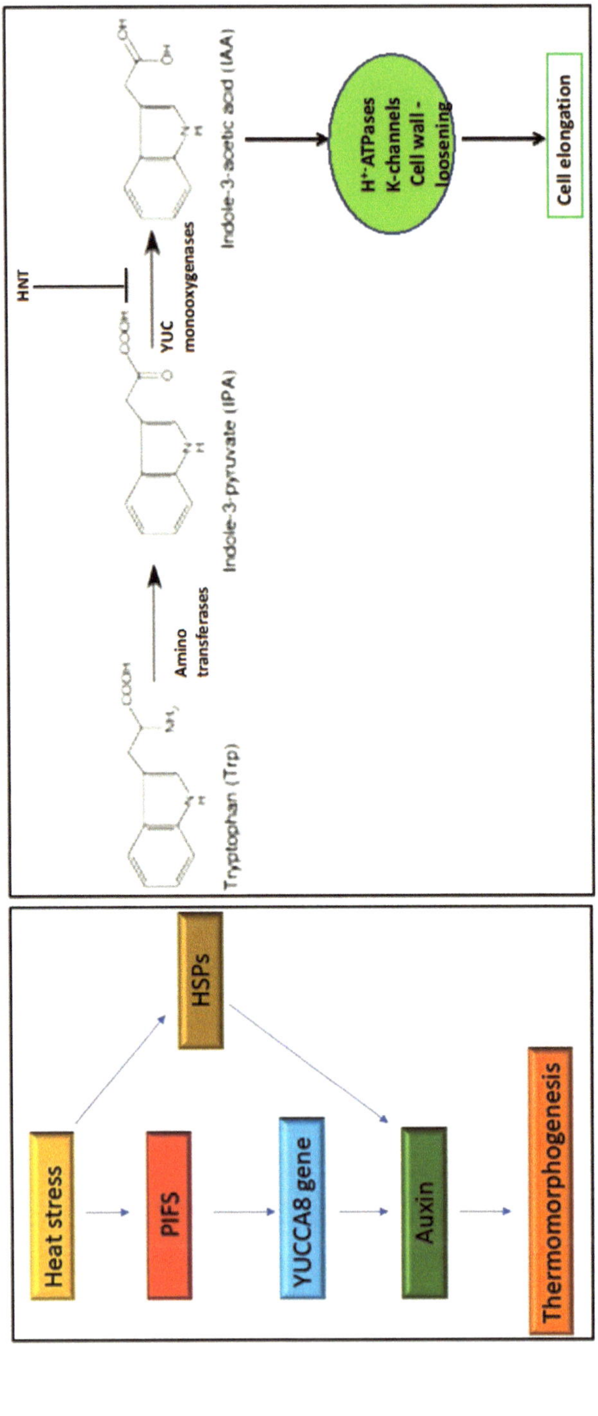

FIGURE 10.3 Molecular mechanism of auxin-induced thermomorphogenesis in HNT-stressed plants.

enzymes, cytokinin oxidase/dehydrogenase (CKX).[29] Sobol et al.[30] reported that heat during summer causes preanthesis abortion in the flower primordia of passion fruit, and that can be controlled by CTK application. CTK activates the expression of HSPs and glycine-rich proteins under heat stress. High night temperatures reduced the activity of enzymes that are involved in the synthesis of cytokinins such as isopentenyl transferases (IPTs), lonely guy (LOG), and cytochrome P450 monooxygenase (CYP735A).[27] As a result of reduced cellular CTK levels, the downstream signaling process stalls and so do biological processes such as cell division, nutrient uptake, shoot initiation, and many other functions that are important for plant growth and development. One of the downstream signaling molecules of cytokinin action is the activation of a histidine kinase (HK), which acts as a CTK receptor. The activated HK phosphorylates and activates histidine-containing phosphotransmitter (HPT) proteins. Activated HPTs are translocated to the nucleus, where they activate the transcription of a response regulator (RR) and a cytokinin response factor (CRF), which in turn regulate various biological processes important to plant growth and development. One of the downstream effective signals is the activation of the homeodomain transcription factor called WUSHEL (WUS), which regulates cell proliferation in the shoot apical meristem (SAM).[27,29] Wu et al.[27] observed that under HNT stress, the expression of genes encoding enzymes of CTK biosynthesis was enhanced. The role of CTK in thermoregulation is not very clear due to the complex crosstalk between CTK and other PGRs. In many developmental processes, CTK and GA have an antagonistic role. In passion fruit, GA treatment triggers floral abortion, whereas CTK application minimizes floral abortion under heat stress. Fleishon et al.[31] suggested that CTK may either reduce endogenous GA levels to improve heat tolerance or just impact downstream pathways of the GA signaling without affecting active GA levels.

10.2.3 ABSCISIC ACID

Abscisic acid (ABA) is implicated as a stress hormone or messenger that regulates various biological functions, including abscission, stomatal opening, seed germination, and dormancy, as well as protein storage in plants. ABA acts by binding to its receptor to initiate a downstream stress-responsive pathway.[32,33] The role of ABA to mitigate heat stress is based on the observations that plants deficient in ABA, or when ABA biosynthesis is inhibited, are affected by heat stress and submissive to heat-induced damage.[34,35] Heat elicits a rapid increase in endogenous cellular ABA levels.[36] The mechanism by which ABA enhances stress tolerance is through the induction of a pathway that enhances cellular hydrogen peroxide to the level that induces antioxidant capacity as well as the induction of heat shock proteins[37,38] (Fig. 10.4). The cellular endogenous level of H_2O_2 increased in plants subjected to the exogenous application of ABA. The strategic mechanism involves the activation of NADPH oxidase gene expression. The increased cellular ROS threshold in turn induces the activation of ROS-scavenging enzymes and antioxidants.[39,40] ABA also activates the ABA-responsive element binding proteins (AREBs), which bind to the heat-stress transcription factor (HSF) and the expression of heat shock proteins.[41] Another mechanism of heat-tolerance activity of ABA in plants during heat stress involves the metabolism of sucrose toward enhancing the energy status of the cell. ABA induces the transcription of sucrose metabolic enzymes such as sucrose synthase and invertase as well as its transporter.[42,43] The effect of ABA differs depending on the life stage of the plant and endogenous concentrations. Toh et al.[44] suggested the inhibitory role of ABA in seed germination under heat stress by upregulating ABA biosynthetic genes like ABA1/ZEP and NCED genes and downregulating GA 20-oxidase genes. Similarly, the exogenous application of ABA also showed enhanced thermotolerance in various crops.[45,46] Kumar et al.[47] reported that ABA improved the chickpea seedling survival rate by increasing the concentration of proline, trehalose, and GB under HNT stress. Glaubitz et al.[48] studied the performance of six cultivars with contesting HNT tolerance and reported that sensitive cultivars had upregulated gene expression of heat-protectant proteins induced by ABA signaling.

10.2.4 GIBBERELLIC ACIDS

Gibberellic acids (GAs) contribute to plant height and growth, and certain studies on wheat have shown that plants exposed to a temperature of 25°C have longer leaves as compared to those exposed to 10°C. The endogenous level of GA was also found to increase, paralleling the increased size of the leaves. Pinthus et al.[49] suggest the increased synthesis of gibberellic acid. However, exposure to higher temperatures may have a reverse effect on plant growth due to the induced misfolding of the ER that interferes with the synthesis pathway of gibberellic acid. To mitigate this negative effect,

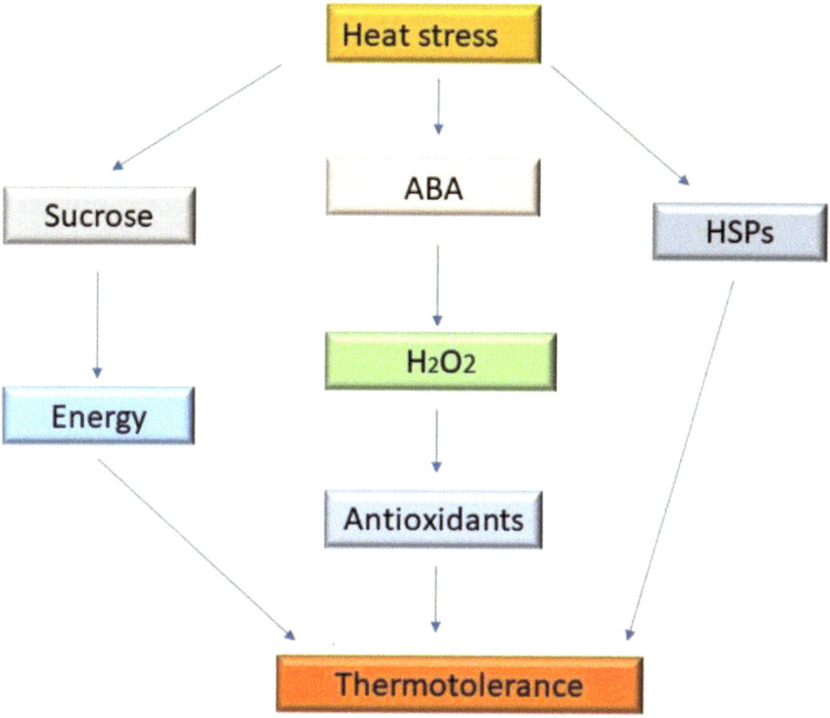

FIGURE 10.4 ABA-mediated thermotolerance in plants under HNT.

exogenous application of PGR, such as triazole, under high temperatures enhances plant growth and development. Triazoles act by inhibiting the enzyme cytochrome P450-dependent monooxygenases, one of the enzymes involved in the synthesis of gibberellic acid present in the ER.[50] Inhibition of the gibberellic synthetic pathway retards plant growth but leads to a compact plant with retarded stem length and a compromised energy expenditure toward shoot growth and more energy is channeled toward flowering, reproduction, and survival[50] (Fig. 10.5).

10.2.5 JASMONIC ACID

Jasmonic acid (JA) is an important plant growth regulator that promotes various processes in the plant such as germination, seedling development, cell cycle regulation, fruit maturation, and the plant defense system. JA is involved in signaling to induce thermotolerance in plants.[12,51] It plays a

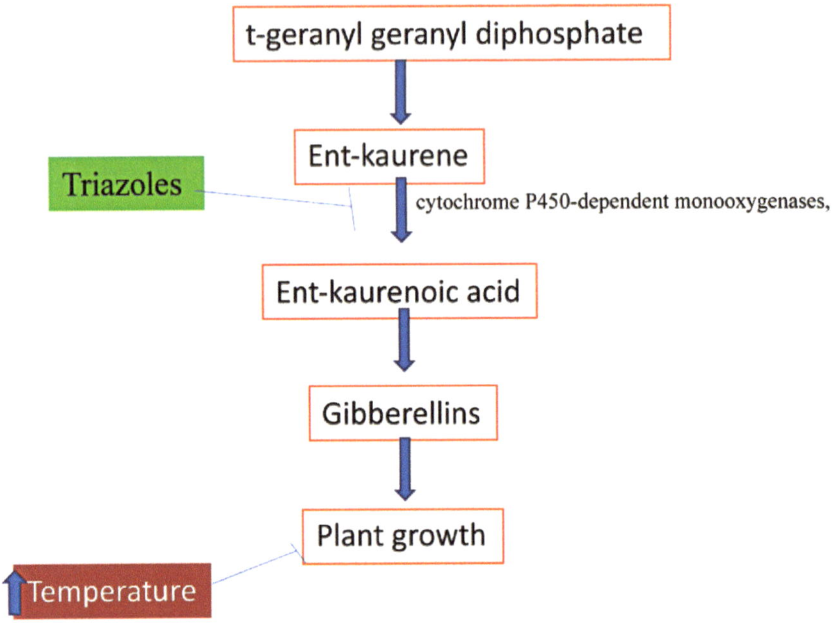

FIGURE 10.5 Molecular mechanism of GA-induced thermotolerance.

major role in signal transmission, and it also stimulates the development of HSPs and other transcription factors that are important for thermotolerance. Wang and Li[52] suggested that JA activates the antioxidant system to mediate protection against heat stress, whereas Clarke et al.[12] suggest that it is due to membrane stabilization. JA is known to play a role in stress, although the underlying processes are not fully understood.[51] Schaarschmidt et al.[53] studied the metabolic profile of six rice cultivars with contrasting HNT tolerance in the leaf and reported that PGRs like SA, JA, and ABA are involved in HNT tolerance. Similar findings were observed by Glaubitz et al.,[48] who identified HNT-tolerant rice cultivars with inositol phosphates, indicating the involvement of JA in HNT response.

10.2.6 *ETHYLENE*

Ethylene (ET) is part of a number of developmental phases, including fruit ripening, growth, and stress reactions.[54] Jin et al.[55] reported thermomorphogeneis mediated by ET at high temperatures by repressing hypocotyl growth. The ethylene-responsive gene (EIN3) is regulated by ethylene-responsive factors (ERFs). Expression of ERF is upregulated following exposure of plants to heat stress. High temperatures induce the degradation of EIN3-binding F-box protein 1/2 (EBF1/2), which is a negative regulator of ERF expression. This is done via a ubiquitin ligase (SIDR1) that leads to inhibition of hypocotyl growth.[56] ERFs are also involved in other PGR signaling pathways under abiotic stresses. Studies in the *A. thaliana* plant revealed that heat stress results in the accumulation of both JA and ET.[37] However, ET and JA antagonistically regulate the plant's response to heat stress.

10.2.7 GLYCINE BETAINE

Another naturally occurring PGR called glycine betaine (GB) acts as an osmolyte and is induced under heat stress conditions, suggesting its protective role for the plant as stress-tolerance[57] (Fig. 10.6). Glycine betaine protects plants by enhancing the expression of stress response genes such as ascorbate oxidase, monodehydroascorbate reductase (MDHAR), catalase (CAT), superoxide dismutase (SOD), and peroxidase (POD). These enzymes scavenge the heat stress-induced increase of reactive oxygen species.[58–60] Another aspect of glycine betaine's role is the activation of mitogen-activated protein kinases and calcium-dependent protein kinases (CDPKs). The activated kinases further activate the downstream signaling molecules, one of which is the heat-shock transcription factor (HSF). The activated HSF enhances the expression of heat-shock proteins, which act to chaperone the proper folding of proteins, thereby preventing heat-induced aggregation of proteins in the endoplasmic reticulum.[61,62] Mohammed et al.[63] investigated the effect of HNT on crop physiology and productivity in rice and reported that exogenous application of GB or SA increased total antioxidant capacity under HNT stress. It also enhanced the production of photosynthates, thereby increasing crop productivity.

10.2.8 SALICYLIC ACID

Salicylic acid (SA), another PGR, plays a critical role as a thermotolerant molecule when plants are exposed to heat stress as evidenced by its increased endogenous level when plants are exposed to heat stress[64] (Fig. 10.7). Pea plants, whose SA biosynthesis was inhibited, were shown to

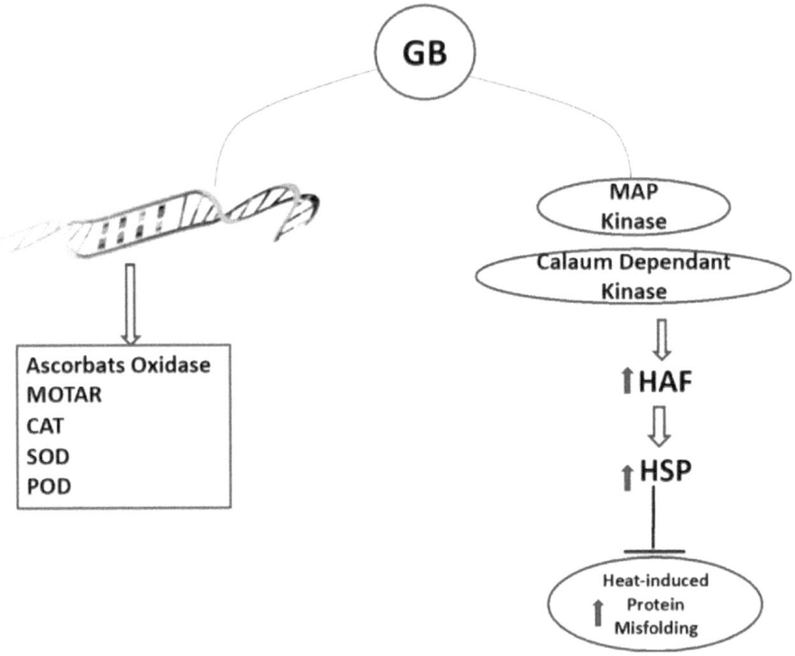

FIGURE 10.6 GB signaling in HNT stress.

have a low tolerance to heat. The mechanism of heat tolerance induction of SA is multifaceted, but the potential signaling involves the activation of nonexpresser of pathogenesis-related (NPR) proteins, enhancing the synthesis of heat shock proteins, Ca^{+2} ATPase activity, and increasing proline oxidase (PROX) activity.[23,65–67]

The high night temperature has a different aspect of the expression of various other plant growth hormones, which is different from the effect imposed by the high day temperature. The differential expression of these hormones is reflected in the morphology of the plants as a whole. Nevertheless, no single PGR acts independently; however, every regulator works in synergy with each other and forms an intricate network that provides an overall response to various environmental stressors. For instance, auxins and cytokinins play an indispensable role in shoot apical meristem development. Auxins promote organogenesis for leaf formation, while cytokinins are for meristem maintenance. Both PGRs function in synergy and are dependent on each other to regulate plant

FIGURE 10.7 Molecular mechanism of thermotolerance by SA under heat stress.

growth and development[29] by helping the plants adjust to the nutrient sink and antioxidant enzyme systems as well as other mechanisms necessary for cell survival.[68,69] In tomato seedlings, the seedlings exposed to high night temperatures show less concentration of gibberellins and auxins, and the seedlings show stunted growth since stem elongation was inhibited, but stem thickness and the number of xylem vessels were normal and not affected when compared to those seedlings exposed to high day temperatures.[23] To regulate seed dormancy and germination, an ideal balance between GA and ABA is needed.

10.2.9 BRASSINOSTEROIDS

Brassinosteroids (BRs) are steroidal plant hormones that are important plant growth regulators. BRs regulate many physiological processes and responses such as flowering, cell division, photosynthesis, growth, and senescence.[27,70] BRASSINAZOLE RESISTANT 1 (BZR1), a BR-regulated transcription factor, plays a major role in BR-mediated thermomorphogenesis.[71] Low BR concentration in the plant under normal temperature causes BZR1 to be phosphorylated and inactivated by protein kinase BRASSINOSTERIOD INSENSITIVE 2 (BIN2). When there is heat stress, BR is accumulated, which inactivates BIN2, leading to the activation of BZR1 by dephosphorylation.

Now BZR1 regulates gene expression and promotes morphological changes for heat stress adaptation (Fig. 10.8). BIN2 also phosphorylates PIR4, which is required for auxin-mediated changes. Accumulated BR also helps in auxin-mediated thermomorphogenesis by deactivating the BIN2 protein.[72,73]

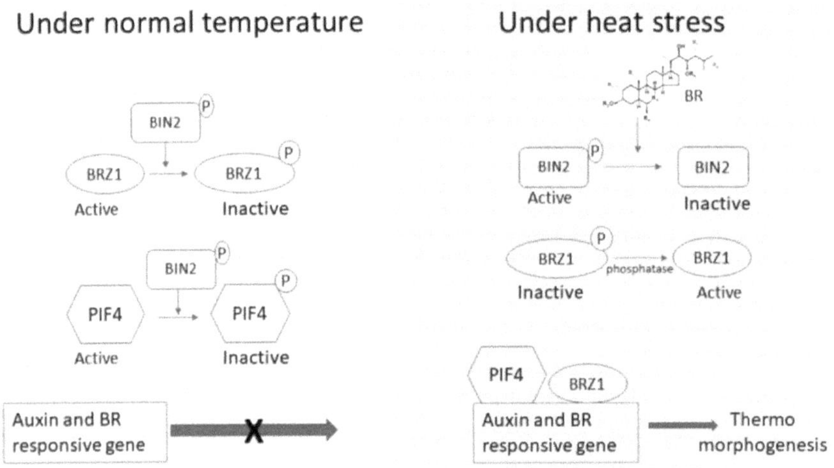

FIGURE 10.8 Molecular mechanism of BR signaling under heat stress.

10.3 CONCLUSIONS

One of the main obstacles limiting crop productivity worldwide is heat stress. Plant growth, photosynthesis, nutrient uptake, and yield are all impacted by heat stress. Studies on how high temperatures affect plant growth and development have primarily looked at daytime temperatures. However, it has been indicated that HNT also has a drastic impact on crop production. To mitigate the detrimental effects of HNT stress, PGRs can be manipulated by breeding techniques, or the exogenous application of various plant growth regulators (PGRs) may help. Over the past few decades, substantial research has been done on the role of PGRs in plants to minimize heat stress, but there have been few studies, particularly looking at the influence of HNT. The development of heat-resistant crops necessitates an understanding of how plants perceive and react to heated temperatures at night. To elucidate PGR function in HNT heat stress, it is necessary to comprehensively investigate the molecular mechanisms underpinning PGR manufacturing and signaling transduction, as well as complex crosstalk across multiple signaling pathways. Given the significance of PGRs in promoting thermotolerance, determining a hormone's role in crop protection from heat stress may have substantial ramifications for breeding programs and plant growth regulator field applications.

KEYWORDS

- **high night temperature**
- **climate change**
- **PGRs**
- **heat stress**
- **HSPs**
- **antioxidant**
- **thermotolerance**
- **thermomorphogenesis**
- **GA**
- **ABA**
- **SA**
- **JA**
- **glycine betaine**

REFERENCES

1. IPCC. (Intergovernmental Panel on Climate Change). *Climate Change: Impacts, Adaptation and Vulnerability. In: Contribution of Working Group II to Fourth Assessment Report of the Intergovernmental Panel on Climate Change*; Parry, M. L., Canziani, O. F., Palutikof, J. P., van der Linden, P. J., Hanson, C. E., Eds.; Cambridge University Press: Cambridge, 2007; p 1000.
2. Alward, R. D.; Detling, J. K.; Milchunas, D. G. Grassland Vegetation Changes and Nocturnal Global Warming. *Science* **1999,** *283*, 229–231.

3. Peng, S.; Huang, J.; Sheehy, J. E.; Laza, R. C.; Visperas, R. M.; Zhong, X.; Centeno, G. S.; Khush, G. S.; Cassman, K. G.; Rice Yields Decline with Higher Night Temperature from Global Warming. *Proc. Natl. Acad. Sci. USA* **2004,** *101*, 9971–9975.
4. Gibson, L. R.; Mullen, R. E. Influence of Day and Night Temperature on Soybean Seed Yield. *Crop Sci.* **1996,** *36*, 98–104.
5. Willits, D. H.; Peet, M. The Effect of Night Temperature on Greenhouse Grown Tomato Yields in Warm Climates. *Agric. For. Met.* **1998,** *92*, 191–202.
6. Izquierdo, N.; Aguirrezábal, L.; Andrade, F.; Pereyra, V. Night Temperature Affects Fatty Acid Composition in Sunflower Oil Depending on the Hybrid and the Phenological Stage. *Field Crops Res.* **2002,** *77*, 115–126.
7. Mohammed, A. R.; Tarpley, L. High Night time Temperatures Affect Rice Productivity through Altered Pollen Germination and Spikelet Fertility. *Agric. For. Met.* **2009,** *149*, 999–1008.
8. Loka, D. A.; Oosterhuis, D. M. Effect of High Night Temperatures on Cotton Respiration. ATP Levels, and Carbohydrate Content. *Environ. Exp. Bot.* **2010,** *68*, 258–263.
9. Bahuguna, R. N.; Jha, J.; Pal, M.; Shah, D.; Lawas, L. M. F.; Sangeetha, K.; Jagadish, S. V. K. Physiological and Biochemical Characterization of NERICA-L-44: A Novel Source of Heat Tolerance at the Vegetative and Reproductive Stages in Rice. *Physiol. Plant* **2015,** *154*, 543–559.
10. Clarke, S. M.; Mur, L. A.; Wood, J. E.; Scott, I. M. Salicylic Acid Dependent Signaling Promotes Basal Thermotolerance But Is Not Essential for Acquired Thermotolerance in *Arabidopsis thaliana*. *Plant J.* **2004,** *38*, 432–447.
11. Miller, G.; Shulaev, V.; Mittler, R. Reactive Oxygen Signaling and Abiotic Stress. *Physiol. Plant.* **2008,** *133*, 481–489.
12. Clarke, S. M.; Cristescu, S. M.; Miersch, O.; Harren, F. J. M.; Wasternack, C.; Mur, L. A. J. Jasmonates Act with Salicylic Acid to Confer Basal Thermotolerance in *Arabidopsis thaliana*. *New Phytol.* **2009,** *182*, 175–187. 10.1111/j.1469-8137.2008.02735.x
13. Liu, J.; Feng, L.; Li, J.; He, Z. Genetic and Epigenetic Control of Plant Heat Responses. *Front. Plant Sci.* **2015,** *6*, 267.
14. Song, L.; Jiang, Y.; Zhao H.; Hou, M. Acquired Thermotolerance in plants. *Plant Cell, Tissue Organ Cult.* **2012,** *111*, 265–276.
15. Sharma, L.; Priya, M.; Kaushal, N.; Bhandhari, K.; Chaudhary, S.; Dhankher, O. P.; Prasad, P. V. V.; Siddique, K. H. M.; Nayyar, H. Plant Growth-Regulating Molecules as Thermoprotectants: Functional Relevance and Prospects for Improving Heat Tolerance in Food Crops. *J. Exp. Bot.* **2020,** *71*, 569–594.
16. Ma, L.; Xia, H.; Meng, Q. Spatiotemporal Variability of Asymmetric Daytime and Night-Time Warming and Its Effects on Vegetation in the Yellow River Basin from 1982 to 2015. *Sensors* **2019,** *19*, 1832.
17. Kupers, J. J.; Oskam, L.; Pierik, R.; Photoreceptors regulate plant developmental plasticity through auxin. *Plants* **2020,** *9*, p.940. DOI: 10.3390/ plants9080940
18. Wang, R.; Zhang, Y.; Kieffer, M.; Yu, H.; Kepinski, S.; Estelle, M. HSP90 Regulates Temperature-Dependent Seedling Growth in Arabidopsis by Stabilizing the Auxin Co-Receptor F-Box Protein TIR1. *Nat. Commun.* **2016,** *7*, 10269. DOI: 10.1038/ ncomms10269.

19. de Wit M, Lorrain, S.; Fankhauser, C. Auxin-Mediated Plant Architectural Changes in Response to Shade and High Temperature. *Physiol. Plant.* **2014,** *151*, 13–24.
20. Oshino, T.; Miura, S.; Kikuchi, S.; Hamada, K.; Yano, K.; Watanabe, M.; Higashitani, A. Auxin Depletion in Barley Plants Under High Temperature Conditions Represses DNA Proliferation in Organelles and Nuclei via Transcriptional Alterations. *Plant Cell Environ.* **2011,** *34*, 284–290.
21. Koini, M. A.; Alvey, L.; Allen, T.; Tilley, C. A.; Harberd, N. P.; Whitelam, G. C.; Franklin, K. A. High Temperature-Mediated Adaptations in Plant Architecture Require the bHLH Transcription Factor PIF4. *Curr. Biol.* **2009,** *19*, 408–413.
22. Sun, J.; Qi, L.; Li, Y.; Chu, J.; Li, C. PIF4-Mediated Activation of YUCCA8 Expression Integrates Temperature into the Auxin Pathway in Regulating *Arabidopsis* Hypocotyl Growth. *PLoS Genet.* **2012,** 8, e1002594. 10.1371/journal.pgen.1002594
23. Reimer, J. J.; Thiele, B.; Biermann, R. T.; Junker-Frohn, L. V.; Wiese-Klinkenberg, A.; Usadel, B.; Wormit A. Tomato Leaves Under Stress: A Comparison of Stress Response to Mild Abiotic Stress Between a Cultivated and a Wild Tomato Species. *Plant Mol. Biol.* **2021,** *107* (3), 177–206.
24. Sakata, T.; Oshino, T.; Miura, S.; Tomabechi, M.; Tsunaga, Y.; Higashitani, N.; Miyazawa, Y.; Takahashi, H.; Watanabe, M.; Higashitani, A. Auxins Reverse Plant Male Sterility Caused by High Temperatures. *Proc. Natl. Acad. Sci. USA* **2010,** *107*, 8569–8574.
25. Majda, M.; Robert, S. The Role of Auxin in Cell Wall Expansion. *Int. J. Mol. Sci.* **2018,** *19* (4), 951.
26. Pavlu, J.; Novak, J.; Koukalova, V.; Luklova, M.; Brzobohatı, B.; Cerny, M. Cytokinin at the Crossroads of Abiotic Stress Signaling Pathways. *Int. J. Mol. Sci.* **2018,** 19, 2450. DOI: 10.3390/ijms19082450
27. Wu, C.; Cui, K.; Wang, W.; Li, Q.; Fahad, S.; Hu, Q. Heat-Induced Cytokinin Transportation and Degradation Are Associated with Reduced Panicle Cytokinin Expression and Fewer Spikelets Per Panicle in Rice. *Front. Plant Sci.* **2017,** 8, 371. DOI: 10.3389/fpls.2017.00371
28. Wang, K.; Zhang, X.; Ervin, E. Antioxidative Responses in Roots and Shoots of Creeping Bentgrass Under High Temperature: Effects of Nitrogen and Cytokinin. *J. Plant Physiol.* **2012,** 169, 492–500. DOI: 10.1016/j.jplph.2011.12.007
29. Wu, W.; Du, K.; Kang, X.; Wei, H. The Diverse Roles of Cytokinins in Regulating Leaf Development. *Hortic Res.* **2021,** *8, 118.* doi.org/10.1038/s41438-021-00558-3
30. Sobol, S.; Chayut, N.; Nave, N.; Kafle, D.; Hegele, M.; Kaminetsky, R. Genetic Variation in Yield Under Hot Ambient Temperatures Spotlights a Role for Cytokinin in Protection of Developing Floral Primordia. *Plant Cell Environ.* **2014,** 37, 643–657. DOI: 10.1111/pce.12184
31. Fleishon, S.; Shani, E.; Ori, N.; Weiss, D. Negative Reciprocal Interactions Between Gibberellin and Cytokinin in Tomato. *New Phytol.* **2011,** *190* (3), 609–617.
32. Sah, S. K.; Reddy, K. R.; Li, J. Abscisic Acid and Abiotic Stress Tolerance in Crop Plants. *Front. Plant Sci.* **2016,** *7*, 571.
33. Vishwakarma, K.; Upadhyay, N.; Kumar, N.; Yadav, G.; Singh, J.; Mishra, R. K.; Kumar, V.; Verma, R.; Upadhyay, R. G.; Pandey, M.; Sharma, S. Abscisic Acid Signaling and Abiotic Stress Tolerance in Plants: A Review on Current Knowledge and Future Prospects. *Front. Plant Sci.* **2017,** *8*, 161. DOI: 10.3389/fpls.2017.00161

34. Ding, W.; Song, L.; Wang, X.; Bi, Y. Effect of Abscisic Acid on Heat Stress Tolerance in the Calli from Two Ecotypes of Phragmites Communis. *Biol. Plant.* **2010,** *54* (4), 607–613. DOI: 10.1007/s10535-010-0110-3
35. Wang, Y.; Chang, H.; Hu, S.; Lu, X.; Yuan, C.; Zhang, C.; Wang, P.; Xiao, W.; Xiao, L.; Xue, G. P.; Guo, X. Plastid Casein Kinase2 Knockout Reduces Abscisic Acid (ABA) Sensitivity, Thermotolerance, and Expression of ABA-and Heat-Stress-Responsive Nuclear Genes. *J. Exp. Bot.* **2014,** *65* (15), 4159–4175.
36. Islam, M.; Feng, B.; Chen, T.; Tao, L.; Fu, G. Role of Abscisic Acid in Thermal Acclimation of Plants. *J. Plant Biol.* **2018,** *61* (5), 255–264.
37. Larkindale, J.; Hall, J. D.; Knight, M. R.; Vierling, E. Heat Stress Phenotypes of Arabidopsis Mutants Implicate Multiple Signaling Pathways in the Acquisition of Thermotolerance. *Plant Physiol.* **2005,** *138*, 882–897. DOI: 10.1104/pp.105.062257
38. Li, N.; Euring, D.; Cha, J. Y.; Lin, Z.; Lu, M.; Huang, L. J.; Kim, W. Y. Plant Hormone-Mediated Regulation of Heat Tolerance in Response to Global Climate Change. *Front. Plant Sci.* **2021,** *11*, 627969.
39. Larkindale, J.; Knight, M. R. Protection Against Heat Stress-Induced Oxidative Damage in Arabidopsis Involves Calcium, Abscisic Acid, Ethylene, and Salicylic Acid. *Plant Physiol.* **2002,** *128*, 682–695. DOI: 10.1104/pp.010320.
40. Kaya, H.; Takeda, S.; Kobayashi, M. J.; Kimura, S.; Iizuka, A.; Imai, A. Comparative Analysis of the Reactive Oxygen Species-Producing Enzymatic Activity of Arabidopsis NADPH Oxidases. *Plant J.* **2019,** *98*, 291–300.
41. Wang, X.; Zhuang, L.; Shi, Y.; Huang, B. Up-Regulation of HSFA2c and HSPs by ABA Contributing to Improved Heat Tolerance in Tall Fescue and Arabidopsis. *Int. J. Mol. Sci.* **2017,** *18*, 1981. DOI: 10.3390/ijms18091981
42. Rezaul, I. M.; Baohua, F.; Tingting, C.; Weimeng, F.; Caixia, Z.; Longxing, T. Abscisic Acid Prevents Pollen Abortion Under High-Temperature Stress by Mediating Sugar Metabolism in Rice Spikelets. *Physiol Plant.* **2019,** *165*, 644–663.
43. Wind, J.; Smeekens, S.; Hanson, J. Sucrose: Metabolite and Signaling Molecule. *Phytochemical* **2010,** *71*, 1610–1614.
44. Toh, S.; Akane, I.; Asuka, W.; Kazumi, N.; Masanori, O.; Yusuke, J.; Atsushi, H. High Temperature-Induced Abscisic Acid Biosynthesis and Its Role in the Inhibition of Gibberellin Action in Arabidopsis Seeds. *Plant Physiol.* **2008,** *146* (3), 1368–1385.
45. Abass, M.; Rajashekar, C. B. Abscisic Acid Accumulation in Leaves and Cultured Cells During Heat Acclimation in Grapes. *Hort. Sci.* **1993,** *28*, 50–52.
46. Robertson, A. J.; Ishikawa, M.; Gusta, L. V.; MacKenzie, S. L. Abscisic Acid-Induced Heat Tolerance in *Bromus inermis* Leyss Cell-Suspension Cultures: Heat-Stable, Abscisic Acid-Responsive Polypeptides in Combination with Sucrose Confer Enhanced Thermostability. *Plant Physiol.* **1994,** *105*, 181–190.
47. Kumar, S.; Kaushal, N.; Nayyar, H.; Gaur, P. Abscisic Acid Induces Heat Tolerance in Chickpea (*Cicer arietinum* L.) Seedlings by Facilitated Accumulation of Osmoprotectants. *Acta Physiol. Plant.* **2012,** *34*, 1651–1658.
48. Glaubitz, U.; Li, X.; Schaedel, S.; Erban, A.; Sulpice, R.; Kopka, J.; Zuther, E.; Integrated analysis of Rice Transcriptomic and Metabolomic Responses to Elevated Night Temperatures Identifies Sensitivity- and Tolerance-Related Profiles. *Plant Cell Environ.* **2017,** *40*, 121–137.

49. Pinthus, M. J.; Gale, M. D.; Appleford, N. E.; Lenton, J. R.; Effect of Temperature on Gibberellin (GA) Responsiveness and on Endogenous GA (1) Content of Tall and Dwarf Wheat Genotypes. *Plant Physiol.* **1989,** *90* (3), 854–859. DOI: 10.1104/pp.90.3.854.
50. Rademacher, W.; Chemical Regulators of Gibberellin Status and Their Application in Plant Production. *Annl. Plant Rev.* **2016,** *49*, 359–403.
51. Sharma, M.; Laxmi, A.; Jasmonates: Emerging Players in Controlling Temperature Stress Tolerance. Front. *Plant S*ci. **2015,** *6*, 1129. DOI: 10.3389/fpls.2015.01129
52. Wang, L.; Li, S. H. Salicylic Acid-Induced Heat or Cold Tolerance in Relation to Ca^{2+} Homeostasis and Antioxidant Systems in Young Grape Plants. *Plant Sci.* **2006,** *170*, 685–694.
53. Schaarschmidt, S.; Glaubitz, U.; Erban, A.; Kopka, J.; Zuther, E.; Differentiation of the High Night Temperature Response in Leaf Segments of Rice Cultivars with Contrasting Tolerance. *Intel. J. Mol. Sci.*, **2021,** *22* (19), 10451.
54. Zhu, Z. Molecular Basis for Jasmonate and Ethylene Signal Interactions in Arabidopsis. *J. Exp. Bot.* **2014,** *65* (20), 5743–5748.
55. Jin, H.; Pang, L.; Fang, S.; Chu, J.; Li, R.; Zhu, Z. High Ambient Temperature Antagonizes Ethylene-Induced Exaggerated Apical Hook Formation in Etiolated Arabidopsis Seedlings. *Plant Cell Environ.* **2018** *41* (12), 2858–2868. doi.org/10.1111/pce.13417
56. Hao, D. D.; Jin, L.; Wen, X.; Yu, F. F.; Xie, Q.; Guo, H. W. The RING E3 ligase SDIR1 Destabilizes EBF1/EBF2 and Modulates the Ethylene Response to Ambient Temperature Fluctuations in Arabidopsis. *Proc. Natl. Acad. Sci. USA* **2021,** *118* (6), e2024592118.
57. Demiral, T.; Türkan, I.; Does Exogenous Glycinebetaine Affect Antioxidative System of Rice Seedlings Under NaCl Treatment? *J. Plant Physiol.* **2004,** *161*, 1089–1100.
58. John, E.; Connolly, E. L. Glycine Betaine Enhances Extracellular Processes Blocking ROS Signaling During Stress. *Plant Signal Behav.* **2009,** *4* (3), 197–199.
59. Annunziata, M. G.; Ciarmiello, L. F.; Woodrow, P.; Dell'Aversana, E.; Carillo, P.; Spatial and Temporal Profile of Glycine Betaine Accumulation in Plants Under Abiotic Stresses. *Front. Plant Sci.* **2019,** *10*, 230.
60. Zulfiqar, F.; Ashraf, M. Bioregulators: Unlocking Their Potential Role in Regulation of the Plant Oxidative Defense System. *Plant Mol. Biol.* **2021,** *105*, 11–41.
61. Divya, K.; Bhatnagar-Mathur, P.; Sharma, K. K.; Reddy, P. S. Heat Shock Proteins (Hsps) Mediated Signalling Pathways During Abiotic Stress Conditions. In *Plant Signaling Molecules*; Woodhead Publishing, 2019; pp 499–516.
62. Andrási, N.; Pettkó-Szandtner, A.; Szabados, L. Diversity of Plant Heat Shock Factors: Regulation, Interactions, and Functions. *J. Exp. Bot.* **2021,** *72* (5), 1558–1575.
63. Mohammed, A. R.; Tarpley, L. Effects of High Night Temperature on Crop Physiology and Productivity: Plant Growth Regulators Provide a Management Option. In: *Global Warming Impacts-Case Studies on the Economy, Human Health, and on Urban and Natural Environments*, 2011; pp 53–172.
64. Pan, Q.; Zhan, J.; Liu, H.; Zhang, J.; Chen, J.; Wen, P.; Huang, W. Salicylic Acid Synthesized by Benzoic Acid 2-Hydroxylase Participates in the Development of Thermotolerance in Pea Plants. *Plant Sci.* **2006,** *171*, 226–233.

65. Rai, K. K.; Pandey, N.; Rai, S. P. Salicylic Acid and Nitric Oxide Signaling in Plant Heat Stress. *Physiol. Plant.* **2020,** *168*, 241–255.
66. Snyman, M.; Cronjé, M. J. Modulation of Heat Shock Factors Accompanies Salicylic Acid-Mediated Potentiation of Hsp70 in Tomato Seedlings. *J. Exp. Bot.* **2008,** *59*, 2125–2132.
67. Khan, M. I. R.; Iqbal, N.; Masood, A.; Per, T. S.; Khan, N. A. Salicylic Acid Alleviates Adverse Effects of Heat Stress on Photosynthesis Through Changes in Proline Production and Ethylene Formation. *Plant Signal Behav.* **2013,** *8* (11), e26374.
68. Ahmad, B.; Zaid, A.; Sadiq, Y.; Bashir, S.; Wani, S. H. Role of Selective Exogenous Elicitors in Plant Responses to Abiotic Stress Tolerance. In: *Plant Abiotic Stress Tolerance*; Hasanuzzaman, M., Hakeem, K. H., Nahar, K., Alharby, H. F., Eds.; Springer: Cham, 2019; pp 273–290.
69. Fahad, S.; Bajwa, A. A.; Nazir, U.; Anjum, S. A.; Farooq, A.; Zohaib, A.; Sadia, S.; Nasim, W.; Adkins, S.; Saud, S.; Ihsan, M. Z.; Alharby, H.; Wu, C.; Wang, D.; Huang, J. Crop Production Under Drought and Heat Stress: Plant Responses and Management Options. *Front. Plant Sci.* **2017,** *8*, 1147. https://doi.org/10.3389/fpls.2017.01147.
70. Bari, R.; Jones, J. D. Role of Plant Hormones in Plant Defence Responses. *Plant Mol. Biol.* **2009,** *69*, 473–488.
71. Ibanez, C.; Delker, C.; Martinez, C.; Bürstenbinder, K.; Janitza, P.; Lippmann, R. Brassinosteroids Dominate Hormonal Regulation of Plant Thermomorphogenesis via BZR1. *Curr. Biol.* **2018,** 28, 303–310. e3. DOI: 10.1016/j.cub.2017.11.077
72. He, J. X.; Gendron, J. M.; Yang, Y. L.; Li, J. M.; Wang, Z. Y. The GSK3-Like Kinase BIN2 Phosphorylates and Destabilizes BZR1, a Positive Regulator of the Brassinosteroid Signaling Pathway in Arabidopsis. *Proc. Natl. Acad. Sci. USA* **2002,** *99* (15), 10185–10190. doi.org/10.1073/pnas.152342599
73. Kim, T. W.; Guan, S.; Sun, Y.; Deng, Z.; Tang, W.; Shang, J. X.; Sun, Y.; Burlingame, A. L.; Wang, Z. Y. Brassinosteroid Signal Transduction from Cell-Surface Receptor Kinases to Nuclear Transcription Factors. *Nat. Cell Biol.* **2009,** *11* (10), 1254–U1233.

CHAPTER 11

Omics Under Elevated Night Temperature

S. G. HARSHA[1,2], RACHANA K. PAWAR[3], ASHA SASTYA[3], SNEHA[4], KARTHIK NANAIAH[3], MAYANK PRATAP SINGH BANGARI[3], H. S. SHEELA[1], and K. S. SHIVASHANKARA[1]

[1]*Division of Basic Sciences, Indian Institute of Horticultural Research, Hesaraghatta Lake Post, Bengaluru, India*

[2]*Department of Plant Physiology, Indira Gandhi Krishi Vishwavidyalaya, Raipur, India*

[3]*Department of Crop Physiology, University of Agricultural Sciences, Bengaluru, India*

[4]*Department of Plant Physiology, Kerala Agricultural University, Vellayani, India*

ABSTRACT

A steady increase in global temperature as an immediate effect of climate change has raised maximum and minimum temperatures, altered day–night temperature balance, and, with minimal chances for radiant heat loss at night, led to elevated night-time temperatures. This unusual rise in night temperature (HNT) appears to negatively influence circadian rhythms, interfering with the normal functioning, growth, and development of a plant. This reduction in plant growth is mainly because of the disturbances at cellular, molecular, and genetic levels causing redox imbalance, fluctuation of enzyme activity, and disruption of biological function. Being

Impact of High Night Temperature on Plant Biology: Toward Sustainable Plant Adaptation to Climate Change. Rajesh Kumar Singhal, Raju Bheemanahalli, Saurabh Pandey, and Pratibha M. D. (Eds.)

sessile and immobile, plants have developed diversified evolutionary patterns that are effective in efficiently stabilizing the ill effects of HNT. To understand plant adaptations under HNT, studying multiple dimensions covering entire plant developmental regulations is necessary. This is where omics techniques such as genomics, transcriptomics, metabolomics, and phenomics come into play for the complete and comprehensive study of internal mechanisms within a plant. Genomics provides complete information about the plant genome, identifies key genes controlling stress tolerance, and helps in their functional analysis and characterization. Transcriptomic analysis reveals the expression levels of genes, their location, function, regulation, interaction, and degradation during stress. Metabolomic profiling provides qualitative and quantitative knowledge about putative metabolites regulating biochemical responses during stress and identifies modifications in metabolic pathways. Phenomics provides the phenotypic variation in plants under stress and helps to know modifications in morphological traits under stress. Combining these techniques in studying the effects of HNT gives a multidimensional view of regulatory mechanisms in plants. Further, these technologies help in understanding the mechanisms governing plant phenotypes under stress and allow us to identify biomarkers for future crop improvement studies.

11.1 INTRODUCTION

Plants are the well-established, highly coordinated, and self-sustaining living systems present on the globe that represent all possible mechanisms to withstand, evolve, and carry forward their generations on this continuously changing planet Earth. Their ability to combat and survive unfavorable environmental conditions provokes us to understand the fascinating science behind their mechanisms to overcome these stressful situations. Until recent times, drought, salinity, flooding, heat, freezing, and heavy metal stress were predominantly studied by our scientists without giving much attention to a plant's response to these stresses during the night. During the past century, global surface temperature has increased by an average of 0.85°C, and it has been predicted to increase by 3.7°C by 2100.[1] Global night temperatures have increased by about 0.204°C per decade (compared to a 0.141°C rise in day temperature) during the 20th century, resulting in a decrease in diurnal temperature amplitude.[2] Increased global temperatures have resulted in an uneven distribution of day–night temperatures, which, as a result, leads to the development of hotter nights.

Many aspects are responsible for an increase in night-time temperatures, such as increased and prolonged cloudy weather limiting the reflection of radiations back into space,[3] frequent ocean heat waves,[4,5] deforestation and desertification,[6] change in land-use pattern for agriculture,[7] increased stubble burning,[8] aerosols,[9] higher evaporation and precipitation,[10] or the differential depth of the planetary boundary layer.[11]

Being immobile organisms, plants have continuously adapted and evolved to these changing environmental abnormalities. Understanding the different mechanisms and strategies plants have followed to withstand these stressful situations helps us in adapting these techniques for crop improvement to meet the needs of growing populations. In recent years, omics-based studies have yielded a wide amount of knowledge about the plant's response to abiotic stresses.[12] Advancements in science and technology have created several omics branches dealing specifically with the internal and molecular components of cellular biology. To date, the major omics approaches include genomics, transcriptomics, proteomics, metabolomics, phenomics, lipidomics, and ionomics.[13] Omics approaches provide a more accurate and precise view of the different molecules governing plants' responses to abiotic stress. Further integration of different omics techniques gives us a multidimensional view of complex biochemical and molecular changes within the plant's body.[14,15]

By planning a proper experimental workflow, keeping these advanced techniques in mind, we can connect all aspects of a plant's phenotypic response as governed by its genotypic regulation mediated by its hub of messengers, metabolites, proteins, lipids, and ions, which gives us a realistic picture of all possible interactions and crossplay within a plant under a given external condition. To make this happen and apply it to our routine research, a thorough knowledge of each of these omics techniques is very important. This chapter gives a brief view of how these omics techniques are currently being used for a better understanding of plants' responses during elevated night temperatures, their advantages over traditional methods, and the different techniques followed under these approaches.

11.2 EFFECT OF ELEVATED NIGHT TEMPERATURES ON PLANTS

One of the most common effects of climate change is high temperatures, which kept us on the edge of a catastrophic collapse in terms of global food security. In recent studies, it has been shown that the increase in night temperature is twice the increase in day temperature. An increase in daytime temperature does not affect a wide geographical area, but on the other hand, an increase in the minimal temperature (night temperature) can last for a longer duration over a wide geographical area.[16] Night temperature impacts plant growth and development more compared to day temperature and poses serious morphological abnormalities that interfere with completing its life cycle.[17]

High night temperatures during the grain-filling stage in wheat and barley reduce the yield by 4%.[18] Narayanan et al.[19] also observed a reduction in yield in wheat due to HNT. The grain-filling stage seems to be a sensitive stage for HNT stress, especially in rice and wheat.[20] HNT in the postflowering and grain-filling stages shortened the grain-filling stage, which accelerates senescence and limits resource transfer from source to sink.[21] This will limit the availability of substrate for starch synthesis and result in poor grain-filling in the case of rice.[22] Earlier studies on heat stress showed a reduction in the activity of starch biosynthesis enzymes and increased the activity of starch-degrading enzymes.[23,24] During the night, respiration and carbohydrate metabolism are highly regulated by temperature. At night, several studies have shown an increase in the rate of respiration when affected by HNT, and the carbon that is getting broken down during this excessive respiration could have been used in grain-filling instead of being a substrate supporting a high rate of respiration, thereby compensating for yield reduction.[16,25] However, HNT during the vegetative stage does not have a significant impact on biomass production and yield.[26,27] This effect can be because the high demand for carbon can be satisfied with active photosynthesis. As in the grain-filling stage, the induction of senescence decreases the activity of several key chloroplast enzymes, which leads to a reduction in the antioxidant capacity, a decrease in the chlorophyll content, and the production of more ROS. This hampers active photosynthesis, which is unable to meet the high demand for carbohydrates due to increased respiration. HNT during the postreproductive stage not only affects yield but also the quality of the final product. The accumulation of stress-related proteins and lipids in the wheat grain affects the qualitative parameters of the grain.[21,28,29]

Moreover, studies done on wheat seedlings under HNT show that it can increase the transpiration rate the following day along with a decrease in leaf area.[30] They further hypothesized that HNT affects hydraulic conductivity by increasing aquaporin activity while negatively affecting photosynthesis, leading to decreased leaf area.

One of the hidden and least exploited effects of HNT is the increase in night-time VPD.[16] As it is well understood, stomata cannot be closed completely at night. They still provide a small opening for the diffusion of gases. Night-time transpiration can play a role in leaf cooling, but can an increase in night-time VPD regulate the stomatal opening? Still, HNT and its increase in VPD at night are not well understood. However, studies have reported that barley,[31] maize,[32] tomato,[33] and grapevine[34,35] lose a significant amount of water at night, which can be driven by VPD.

Elevated night temperatures usually lead to several physiological disorders and abnormalities and disturb the functional behavior of plants. High temperatures induce pollen sterility, delay anther dehiscence, poor pollination, reduced pollen germination on the stigma and pollen tube elongation, and impair grain-filling, resulting in poor seed set, yield loss, and degradation of grain quality in cereals and pulses.[36,37] The warmer the night temperature, the faster the respiration processes, limiting the number of sugars and other storage products that accumulate in fruits and developing seeds,[38] which induce bolting and flower drop[39] in many vegetable crops.

11.3 OMICS APPROACHES FOR UNDERSTANDING ELEVATED NIGHT TEMPERATURES

Recent developments and progress in the fields of science, technology, artificial intelligence, and bioinformatics have led to the origin of a new field of study commonly termed omics approaches. With the advent of omics technologies, tremendous progress has been made in understanding the signaling and molecular pathways in normal and stressed organisms.[40] Omics refers to the study of large sets of molecules in biology for the understanding of genes (genomics), genomes (epigenomics), mRNAs (transcriptomics), noncoding RNA transcripts (miRNAomics), proteins (proteomics), metabolites (metabolomics), lipids (lipidomics), and ions (ionomics). Applying these techniques either individually or in combination allows for advancing our understanding of biological processes within a short period of time. Integration of these omics techniques helps to globally measure the different classes of cellular biomolecules present in an organism and enables a systematic study of dynamic molecular processes that regulate cell functions like growth, sensing, and environmental adaptation.[41] Some of the most widely used omics techniques for understanding HNT stress are discussed below and highlighted in Figure 11.1.

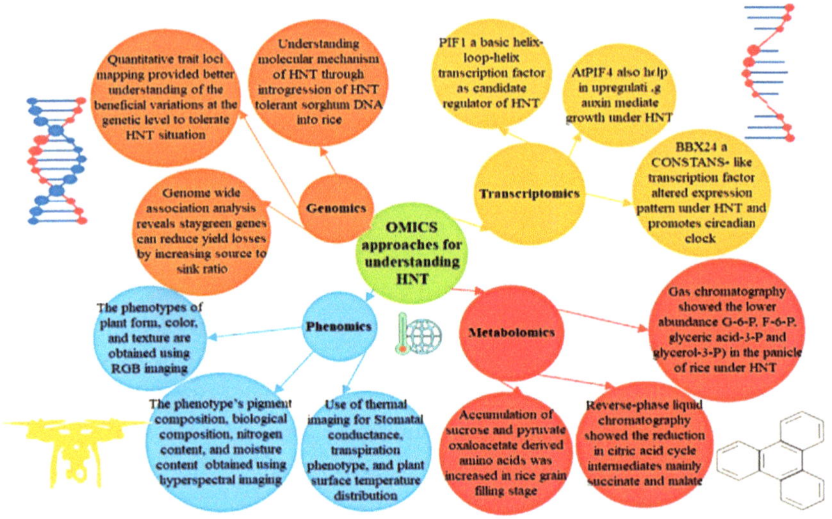

FIGURE 11.1 Graphical representation of omics approaches for understanding nightwarming responses in plants.

11.3.1 GENOMICS

Understanding the basic genetic architecture of the plant is the need of the hour to address the upcoming adverse climate by developing and releasing tolerant varieties. Nowadays, there are highly advanced techniques in bioinformatics to analyze the huge amounts of DNA that constitute the genome of an organism, that is, Genomics, the study of all the genes present in an organism that constitute its genome.

Recently, genome-wide association analysis and network analysis of genes under high night temperatures (HNTs) ~29°C have revealed that genes associated with stay greenness might be involved in reducing yield losses by increasing the efficiency of the source-to-sink ratio during the grain-filling process in rice *indica* genotypes.[42]

In addition, genomic regions linked to the particular trait of choice were evaluated by quantitative trait loci (QTL) mapping in the recombinant inbred line (RIL) population obtained from HNT-tolerant and sensitive varieties in the U.S. Results have provided a better understanding of the beneficial variations at the genetic level to tolerate stressful situations.[43]

There are various tools available to screen for genes and variations in the genomic regions. Zhou et al.[44] have used a hybrid sequencing strategy to analyze differentially expressed genes (DEGs) and single nucleotide polymorphisms (SNPs) associated with HNT. Results indicate the possible activation of specific physiological processes like cellular respiration, disturbed Reactive Oxygen Species (ROS) homeostasis, and reduced energy efficiency in the susceptible lines compared to tolerant lines. These serve as important bioresources for breeding tolerant cultivars.

Xu et al.[45] explored the MAGIC population for several physiological and developmental events in rice. GWAS analysis revealed a negative relationship between the nonstructural carbohydrates and yield, especially during the night period in the stem region. Further candidate genes, an invertase inhibitor and UDP-glycosyl transferase, are known to be negatively involved in HNT conditions. Zhang et al.[46] experimented to understand the molecular mechanism by a comparative genomics approach, wherein they introgressed HNT-tolerant sorghum DNA into rice. By comparative genome analysis, they reported repetitive element variation in the copy number in the recipient lines. This serves as base information to further understand the molecular mechanism and to design a breeding strategy.

For thorough genetic analysis, QTL mapping is widely used to map the chromosomes using different molecular markers such as RFLP, SNP, and SSR using different mapping populations such as F2, BC, RILS, NILS, DH, etc.[47]

The genomic study provides insights, basic information, and a leading direction to develop and release the tolerant cultivars suitable for future warm climatic conditions. These serve as crucial bioresources to improve crop productivity.

11.3.2 TRANSCRIPTOMICS

The transcriptome refers to the entire RNA transcripts present in a specific cell or tissue at a specific time point under a specific physiological condition, including mRNA, tRNA, rRNA, and other noncoding RNAs.[48] Transcriptomics deals with the study of gene expression at the RNA level to provide information on gene structure and functions regulating the molecular mechanisms of a biological system.[49] The complex response of plants against HNT can have a temporal and spatial difference in gene expression profiling. With the advancement in sequencing technologies, transcriptomics has evolved from a chip hybridization platform to RNA sequencing technology (RNA-seq).[50]

This can be useful to perform transcriptomic analysis without a genomic reference. Transcriptome profiling can contribute to an in-depth understanding of plant responses toward the HNT. Transcriptomic analysis complemented with NGS and RNA-seq for sRNA can improve the plant's genomic resources. Transcriptomics studies can help fill the gap between the molecular mechanism and yield loss due to HNT. The asymmetric warming between day and night is a critical variable to be considered. HNT is a very complex phenomenon; even slight changes in day and night temperature can affect a lot of molecular and cellular activities.

Studying alone the effect of HNT cannot give a reasonable answer to hiding behind warm curtains. Reports have suggested that HNT can affect the transcript level during the day.[51] Transcriptomic studies should be focused on thermoperiodic gene expression profiling. Transpiration, respiration, and photosynthesis are thermosensitive processes. But only respiration occurs consistently at both night and day. HNT promotes a higher respiration rate. At night, light-driven photoprotective mechanisms are absent, which promotes the production of ROS during the night due to HNT.

Desai et al.[51] identify PIF1, a basic helix-loop-helix transcription factor, as a candidate regulator of HNT. In arabidopsis, AtPIF4 and AtPIF5 are known to interact with circadian clock elements.[52] AtPIF4 also helps in upregulating auxin and mediates growth at high temperatures.[53] Moreover, under high temperatures, AtPIF4 is regulated by the AtPHYB. AtPHYB interacts with AtELF3, which is part of the evening complex, a core component of the plant circadian clock.[52,54] These findings indicate the possible role of HNT in the plant circadian clock as well. The plant circadian clock contributes to carbon homeostasis, and disruption in carbon homeostasis due to the HNT can affect carbon mobilization as well.[51]

BBX24, a CONSTANS-like transcription factor, also identified as a DEG under HNT,[51] is an activator of PIF activity and involves the photoperiodic regulation of flowering.[55,56] Arabidopsis BBX24 was also identified as a target of the evening complex. AtELF4, a thermosensor, also shows the altered expression pattern under HNT, which is also known to promote the circadian clock in the absence of the day/night cycle.[51,57]

In next-generation sequencing technologies, PCR amplification, template migration, base mismatch, and GC preferences are the common issues, which can affect the accuracy of sequence results.[58] PacBio single-molecule real-time (SMRT) sequencing technology and Oxford nanopore sequencing technology are the two main next-generation sequencing technologies. Moreover, the SMRT sequencing technology is based on the idea of synthesis sequencing. SMRT technology offers two distinct advantages over previous approaches. To solve the issue of background noise, the fluorescent group is first linked to the phosphoric acid group. Second, amplification is not required, and SMRT allows for more precise measurement by self-correction. Moreover, nanopore single-molecule sequencing technology is a next-generation single-molecule real-time sequencing device that estimates base composition mostly through changes in electrical impulses. The advantages of nanopore sequencing are high throughput, extended read length, and low cost. It does, however, have several downsides, such as a high single-base mistake rate and random error.[58]

11.3.3 METABOLOMICS

The plant responds to the outer environment by producing a wide variety of metabolites, which are varied under stress to overcome their negative effects. Metabolites are the final products of different regulatory processes such as transcription, translation, and gene regulation. Thus, the metabolome in response to stress could more closely reflect the molecular phenotype of the plant, giving a better view of the internal mechanisms that provide tolerance against stress.[59] When a plant is exposed to high night temperatures, it imparts negative effects on various physiological and biochemical processes, such as photosynthesis, respiration, primary and secondary metabolism, water relations, lipid and protein metabolism, and activates several signaling cascades and complex crosstalk between different signaling pathways involving hormones, kinases, phosphatases, phenols, and many other compounds.[60]

Many analytical techniques, like gas chromatography tandem mass spectrometry (GC-MS), liquid chromatography tandem mass spectrometry (LC-MS), capillary electrophoresis tandem mass spectrometry (CE-MS), high performance liquid chromatography (HPLC), UPLC (ultra performance liquid chromatography), AP-MS (affinity purification coupled to mass spectrometry), NMR (nuclear magnetic resonance), SPMESH-GC-MS (solid-phase mesh-enhanced sorption from headspace coupled to GC–MS), SFC (supercritical fluid chromatography), time-of-flight (TOF),

Fourier transfer ion cyclotron resonance (FT-ICR), hydrophilic interaction liquid chromatography (HILIC), reversed-phase liquid chromatography (RPLC), immobilized metal affinity chromatography (IMAC), ion mobility spectrometry (IMS), INST-MFA (isotopically nonstationary metabolic flux analysis), triple quadrupole mass spectrometry (QqQ-MS) and quadrupole-linear ion-trap mass spectrometry (QLIT-MS), mass spectrometry imaging (MSI), laser ablation electrospray ionization (LAESI), matrix-assisted laser desorption ionization (MALDI), etc., are being presently used to understand the metabolomic variation in plants either in targeted or untargeted way.[61] A combination of these techniques helps in getting more precise and accurate predictions of key metabolites.

The grain yield of rice is influenced by carbon and nitrogen supply during the grain-filling stage, and this stage is very much affected by HNT. The carbon source is lost to compensate for temperature-sensitive respiration, which was confirmed by the metabolome profiling by gas chromatography coupled to electron impact ionization-time of flight mass spectrometry (GC/EI-TOF-MS), showing the lower abundance of sucrose and other glycolytic pathway intermediates (glucose-6-phosphate, fructose-6-phosphate, glyceric acid-3-phosphate, and glycerol-3-phosphate) in the panicle of rice plant exposed to HNT.[62–64] HNT also impairs photosynthesis and abrupt carbohydrate pools in leaves, thereby limiting carbon supply to developing grains and resulting in lesser yields with poor-quality grains.[63] Metabolome profiling of *Pinus radiata* under elevated temperatures by GC-MS and LC-Orbitrap-MS helped in the quantification of differential metabolites, among which cytokinin, flavonoid and terpenoid biosynthesis, and fatty acid metabolism are the most important pathways involved in the heat-stress response, which are coordinated by zeatin riboside and isopentenyl adenosine during multiple and complex interactions.[65]

Picolitre pressure-probe-electrospray-ionization mass spectrometry (picoPPESI-MS) is the latest metabolomic technique that can be applied to directly determine the differentially synthesized metabolites in intact pollen grains to know the effect of HNT on flowering, anthesis, and pollen grain germination.[66] Liquid chromatography–high-resolution mass spectrometry (LC-HRMS) analysis of *Arabidopsis* under heat stress provided information about possible changes in amino acid and sugar metabolism, glycolysis, TCA cycle, GABA shunt, urea cycle, glutathione, and purine metabolism, which altogether help in defense against heat stress.[67]

Ultrahigh-performance liquid chromatography-MS/MS (UHPLC-MS/MS) is also useful to understand the differential metabolites during heat stress in vegetables.[68]

Reverse-phase liquid chromatography-electrospray ionization mass spectroscopy (RPLC-MS)-based nontargeted metabolite studies under HNT in maize showed that it does not have much effect on sugar levels in grain endosperm, whereas citric acid cycle intermediates, mainly succinate and malate, were severely reduced.[69] CE-MS and CE-DAD-based metabolomic analyses to determine the effect of HNT on grain-filling in rice revealed that the accumulation of sucrose and pyruvate oxaloacetate-derived amino acids was increased, and levels of sugar phosphates and organic acids involved in glycolysis and the tricarboxylic acid (TCA) cycle were decreased, respectively.[23] Further, they found that starch deposition is impaired and sucrose import and starch biosynthesis is reduced during HNT stress.

By combining different techniques of metabolite analysis, it is possible to better understand the metabolic pathways that are influenced by HNT and how the plant makes the necessary changes in its metabolite hub to withstand the deleterious effects of HNT stress. Further, combining metabolomics with transcriptomics gives confirmatory evidence on metabolite expression, function, and regulation during HNT stress in plants, and knowing this will help us in planning proper breeding and biotechnological approaches to achieve crop improvement to cope with HNT.

11.3.4 PHENOMICS

Phenotyping is crucial for understanding various physiological processes as well as the responses of plants to their environment. Screening a large number of plants, be they germplasm lines or mapping populations, for specific traits or responses requires high-throughput phenotyping techniques that are precise without compromising the quality of the observation.

Several phenotyping systems are in use that can capture the observations from cellular to whole plant levels. Crop phenotypes are quite complex and vary with the genotype and environment. A phenomics facility could provide a uniform growing condition, up to a certain degree, which helps to study the genetic variability within or among the species for specific traits or responses. A field-based phenomics platform helps capture the observations at the whole plant level. These plants could be grown in pots or in the field. Large gantries fitted with sensors or cameras help collect information noninvasively. The phenotypes of plant form, color, and texture are obtained using RGB imaging. Imaging chlorophyll fluorescence identifies photosynthetic phenotypes. The phenotype's pigment composition, biological composition, nitrogen content, and moisture content are all obtained using hyperspectral imaging. Stomatal conductance, transpiration phenotype, and plant surface temperature distribution are all obtained using thermal imaging. Lidar measures a plant's three-dimensional phenotype. These systems could help accelerate the screening process. Recently, unmanned aerial vehicles (UAVs) equipped with sensors are being put to test to replace ground-based phenotyping systems.[70–72]

High-throughput phenotyping facilities have been successfully used in phenotyping for drought stress. A mini-lysimeter-based phenomics facility has been used to study the acquired tolerance traits of rapid and gradually stressed plants. The response to stress within (rice) and among (rice and wheat) species has been studied by providing a uniform environment while mimicking the natural environment. This gravimetry-based system allowed us to maintain the moisture content of pots at a predetermined level.[73] Similarly, the effects of heat and drought, individually and in combination, have been studied in Brassica genotypes using a high-resolution plant phenomics center.[74]

Using a range of nondestructive plant phenomics tools, such as whole plant imaging, leaf stomatal conductance, thermal imaging, the photochemical reflectance index, the quantum yield of photosynthesis, and leaf gas exchange, these genotypes were studied for tolerance at the anthesis stage. These facilities provide the advantage of collecting data throughout the cropping period. This data is available to be analyzed at any point of time during or after the cropping season. Hein et al.[75] developed field-based heat tents integrating cyber-physical technology to impose high night-time temperatures on 12 genotypes of winter wheat. An average decrease in the yield of about 20.3%, compared to the control, was observed in these genotypes. This system was upgraded to a field-based facility, where the impact of high night temperatures, postanthesis, on the yield of winter wheat was captured.[76]

A few limitations of the phenomics facilities include the huge investment in setting up the facility and trained staff for the running and maintenance of these facilities. These facilities produce a large amount of data.

Machine learning and adopting new models could help reduce the stress of interpretation; however, this needs standardization and validation.[70,72,77] The integration of plant management, engineering, and data science is essential for the success of a phenomics facility.[70,72,77]

Genotyping and phenotyping of appropriate mapping populations are necessary to enable quantitative trait locus (QTL) mapping and genome-wide association studies (GWAS), which have been extensively conducted in crop plants. Integrating largescale QTL or GWAS analysis with high-throughput phenotyping technologies dramatically increased our understanding of the dynamic development process of crops and gave us a brand-new instrument for studying crop genomics, gene characterization, and breeding.[70,71,78]

11.4 CONCLUSIONS AND FUTURE PERSPECTIVES

The average world temperature is expected to rise by 1.0–3.7°C by 2100 (IPCC 2018), which will increase the frequency and severity of elevated temperatures in the tropical and subtropical regions where crops are grown. This forecast has created a new window of rapidly rising night-time temperatures relative to daytime temperatures on a greater scale, posing a significant challenge to the sustainable production of numerous crops and endangering the security of the world's food supply. Crop breeding and improvement projects now prioritize minimizing yield and quality loss brought on by high night temperatures while also increasing crops' tolerance to such temperatures. One such approach is the study of various trait analyses in correlation with the phenotype that explains the underlying physiological responses that are essential for crop improvement and tolerance. Stress index-based mapping of genetic loci suggests that the quantification of stress tolerance in rice against high night temperatures by combining both relative and absolute phenotypes helps to identify accessions with greater productivity and tolerance.[42]

Furthermore, a biochemical approach can be used to study the effect of high night-time temperatures on the quality of harvested produce. The anthocyanin content in the grapes grown at high night-time temperatures was found to be lower due to the reduced activity of anthocyanin biosynthesis genes.[79] Physiological changes include the disturbed source-to-sink relation due to the negative effect of elevated night temperature on pollen fertility and seed set.[80] Heat stress does not cause passive damage to plants;

instead, plants adjust their mRNA, protein, and metabolite composition in response to temperature changes. These adjustments are made to create a new steady-state balance of biological processes that will allow the organism to function, survive, and even procreate at a higher temperature.

Recent thorough investigations of the transcriptome, proteome, and metabolome—aided by information from the rice genome—have identified the fundamental metabolic mechanisms that contribute to the degradation of the grain quality of important crops at high temperatures. Recent advances in science have transformed how plant biologists investigate stress-responsive systems and pathways over the past 10 years.

To develop a comprehensive understanding of stress signaling, research is still needed to examine a variety of aspects of high-night temperature stress responses in plants. Besides having a new insight into the effect of elevated night temperature stress on the major physiological and molecular processes of the crop, this review provides a trajectory of various approaches to improve crop sustainability and tolerance to the ever-changing climatic conditions.

KEYWORDS

- **night warming**
- **growth and development**
- **multi-omics**
- **next-generation sequencing**
- **transcription factor**
- **genotyping**

REFERENCES

1. IPCC. *AR5 Climate Change. Impacts, Adaptation, and Vulnerability*; Cambridge University Press: Cambridge, 2014.
2. Wang, K.; Li, Y.; Wang, Y.; Yang, X. On the Asymmetry of the Urban Daily Air Temperature Cycle. *J. Geophys. Res.: Atmos.* **2017,** *122* (11), 5625–5635.
3. Huang, Y.; Dickinson, R. E.; Chameides, W. L. Impact of Aerosol Indirect Effect on Surface Temperature Over East Asia. *Proc. Natl. Acad. Sci.* **2006,** *103* (12), 4371–4376.

4. Zhongming, Z.; Linong, L.; Xiaona, Y.; Wangqiang, Z.; Wei, L. Climate Change, Impacts and Vulnerability in Europe, 2012.
5. NOAA (National Oceanic and Atmospheric Administration). Heat Stress Datasets and Documentation. Provided to EPA by NOAA in February, 2022.
6. Lawrence, D.; Coe, M.; Walker, W.; Verchot, L.; Vandecar, K. The Unseen Effects of Deforestation: Biophysical Effects on Climate. *Front. Fore. Glob. Chang.* **2022,** 49.
7. Gallo, K. P.; Easterling, D. R.; Peterson, T. C. The Influence of Land Use/Land Cover on Climatological Values of the Diurnal Temperature Range. *J. Clim.* **1996,** 2941–2944.
8. Saxena, P.; Sonwani, S.; Srivastava, A.; Jain, M.; Srivastava, A.; Bharti, A.; Rangra, D.; Mongia, N.; Tejan, S.; Bhardwaj, S. Impact of Crop Residue Burning in Haryana on the Air Quality of Delhi, India. *Heliyon* **2021,** *7* (5), 06973.
9. Ramanathan, V.; Ramana, M. V.; Roberts, G.; Kim, D.; Corrigan, C.; Chung, C.; Winker, D. Warming Trends in Asia Amplified by Brown Cloud Solar Absorption. *Nature* **2007,** *448* (7153), 575–578.
10. Dai, A.; Fung, I. Y.; Del Genio, A. D. Surface Observed Global Land Precipitation Variations During 1900–88. *J. Clim.* **1997,** *10* (11), 2943–2962.
11. Davy, R.; Esau, I.; Chernokulsky, A.; Outten, S.; Zilitinkevich, S. Diurnal Asymmetry to the Observed Global Warming. *Int. J. Climatol.* **2017,** *37* (1), 79–93.
12. Soda, N.; Wallace, S.; Karan, R. Omics Study for Abiotic Stress Responses in Plants. *Adv. Plants Agric. Res.* **2015,** *2* (1), 00037.
13. Shah, T.; Xu, J.; Zou, X.; Cheng, Y.; Nasir, M.; Zhang, X. Omics Approaches for Engineering Wheat Production Under Abiotic Stresses. *Int. J. Mol. Sci.* **2018,** *19* (8), 2390.
14. Chaudhary, J.; Khatri, P.; Singla, P.; Kumawat, S.; Kumari, A.; Vikram, A.; Jindal, S. K.; Kardile, H.; Kumar, R.; Sonah, H.; Deshmukh, R. Advances in Omics Approaches for Abiotic Stress Tolerance in Tomato. *Biology* **2019,** *8* (4), 90.
15. Pirzadah, T. B.; Malik, B.; Hakeem, K. R. Integration of "Omic" Approaches to Unravel the Heavy Metal Tolerance in Plants. *Essent. Bioinform.***2019**, *3*, 79–92.
16. Sadok, W.; Jagadish, S. K. The Hidden Costs of Night-time Warming on Yields. *Trend. Plant Sci.* **2020,** *25* (7), 644–651.
17. Peng, S.; Huang, J.; Sheehy, J.; Laza, R.; Visperas, R.; Zhong, X.; Centeno, G.; Khush, G.; Cassman, K. Rice Yields Decline with Higher Night Temperature from Global Warming. *Proc. Natl. Acad. Sci. USA.* **2004,** *101* (27), 9971–9975.
18. Garcia, G. A.; Serrago, R. A.; Dreccer, M. F.; Miralles, D. J. Postanthesis Warm Nights Reduce Grain Weight in Field-Grown Wheat and Barley. *Field Crops Res.* **2016,** *195*, 50–69.
19. Narayanan, S.; Prasad, P. V. V.; Fritz, A. K.; Boyle, D. L.; Gill, B. S. Impact of High Night-Time and High Daytime Temperature Stress on Winter Wheat. *J. Agron. Crop Sci.* **2015,** *201*, 206–218.
20. Impa, S. M.; Raju, B.; Hein, N. T.; Sandhu, J.; Prasad, P. V.; Walia, H.; Jagadish, S. K. High Night Temperature Effects on Wheat and Rice: Current Status and Way Forward. *Plant Cell Environ.* **2021,** *44* (7), 2049–2065.
21. Impa, S. M.; Vennapusa, A. R.; Bheemanahalli, R.; Sebela, D.; Boyle, D.; Walia, H.; Jagadish, S. V. K. High Night Temperature Induced Changes in Grain Starch

Metabolism Alters Starch, Protein and Lipid Accumulation in Winter Wheat. *Plant Cell Environ.* **2020,** *43*, 431–447.
22. Bahuguna, A.; Khan, I.; Bajpai, V. K.; Kang, S. C. MTT Assay to Evaluate the Cytotoxic Potential of a drug. *Bangladesh J. Pharm.* **2017,** *12* (2), 115–118.
23. Yamakawa, H.; Hakata, M. Atlas of Rice Grain Filling-Related Metabolism Under High Temperature: Joint Analysis of Metabolome and Transcriptome Demonstrated Inhibition of Starch Accumulation and Induction of Amino Acid Accumulation. *Plant Cell Physiol.* **2010,** *51* (5), 795–809.
24. Hakata, M.; Kuroda, M.; Miyashita, T.; Yamaguchi, T.; Kojima, M.; Sakakibara, H.; Yamakawa, H. Suppression of α-Amylase Genes Improves Quality of Rice Grain Ripened Under High Temperature. *Plant Biotechnol. J.* **2012,** *10* (9), 1110–1117.
25. Shi, W.; Yin, X.; Struik, P. C.; Xie, F.; Schmidt, R. C.; Jagadish, K. S. Grain Yield and Quality Responses of Tropical Hybrid Rice to High Night-Time Temperature. *Field Crops Res.* **2016,** *190*, 18–25.
26. Kanno, K.; Mae, T.; Makino, A. High Night Temperature Stimulates Photosynthesis, Biomass Production and Growth During the Vegetative Stage of Rice Plants. *Soil Sci. Plant Nutr.* **2009,** *55* (1), 124–131.
27. Glaubitz, U.; Li, X.; Köhl, K. I.; van Dongen, J. T.; Hincha, D. K.; Zuther, E.; Differential Physiological Responses of Different Rice (*Oryza sativa*) Cultivars to Elevated Night Temperature During Vegetative Growth. *Funct. Plant Biol.* **2013,** *41* (4), 437–448.
28. Dupont, S.; Morsut, L.; Aragona, M.; Enzo, E.; Giulitti, S.; Cordenonsi, M.; Piccolo, S. Role of YAP/TAZ in Mechanotransduction. *Nature* **2011,** *474* (7350), 179–183.
29. Hurkman, W. J.; Vensel, W. H.; Tanaka, C. K.; Whitehand, L.; Altenbach, S. B.; Effect of High Temperature on Albumin and Globulin Accumulation in the Endosperm Proteome of the Developing Wheat Grain. *J. Cereal Sci.* **2009,** *49* (1), 12–23.
30. Schoppach, R.; Sadok, W. Transpiration Sensitivities to Evaporative Demand and Leaf Areas Vary with Night and Day Warming Regimes Among Wheat Genotypes. *Funct. Plant Biol.* **2013,** *40*, 708–718.
31. Even, M.; Sabo, M.; Meng, D.; Kreszies, T.; Schreiber, L.; Fricke, W. Night-Time Transpiration in Barley (*Hordeum vulgare*) Facilitates Respiratory Carbon Dioxide Release and Is Regulated During Salt Stress. *Ann. Bot.* **2018,** *122*, 569–582.
32. Tamang, B. G.; Sadok, W. Nightly Business: Links Between Daytime Canopy Conductance, Nocturnal Transpiration and Its Circadian Control Illuminate Physiological Trade-Offs in Maize. *Environ. Exp. Bot.* **2018,** *148*, 192–202.
33. Caird, M. A.; Richards, J. H.; Hsiao, T. C. Significant Transpirational Water Loss Occurs Throughout the Night in Field-Grown Tomato. *Funct. Plant Biol.* **2007,** *34*, 172–177.
34. Fuentes, S.; Mahadevan, M.; Bonada, M.; Skewes, M. A.; Cox, J. W. Night-time Sap Flow Is Parabolically Linked to Midday Water Potential for Field-Grown Almond Trees. *Irrig. Sci.* **2013,** *31*, 1265–1276.
35. Rogiers, S. Y.; Greer, D. H.; Hutton, R. J.; Landsberg, J. J. Does Night-Time Transpiration Contribute to Anisohydric Behaviour in a *Vitis vinifera* Cultivar. *J. Exp. Bot.* **2009,** *60*, 3751–3763.
36. Wassmann, R.; Jagadish, S. V.; Heuer, S.; Ismail, A.; Redona, E.; Serraj, R.; Singh, R. K.; Howell, G.; Pathak, H.; Sumfleth, K. Climate Change Affecting Rice Production:

The Physiological and Agronomic Basis for Possible Adaptation Strategies. *Adv. Agron.* **2009,** *1* (101), 59–122.
37. Coast, O.; Ellis, R.H.; Murdoch, A.J.; Quiñones, C.; Jagadish, K.S. High Night Temperature Induces Contrasting Responses for Spikelet Fertility, Spikelet Tissue Temperature, Flowering Characteristics and Grain Quality in Rice. *Funct. Plant Biol.* **2014,** *42* (2), 149–161.
38. Johnson, G. How High Heat Affects Vegetables and Other Crop Plants. In: *Weekly Crop Report, Cooperative Extension*; University of Delaware: Newark, DE, 2011.
39. Jeong, S. W.; Kim, G. S.; Lee, W. S.; Kim, Y. H.; Kang, N. J.; Jin, J. S.; Lee, G. M.; Kim, S. T.; Abd El-Aty, A. M.; Shim, J. H.; Shin, S. C. The Effects of Different Night-Time Temperatures and Cultivation Durations on the Polyphenolic Contents of Lettuce: Application of Principal Component Analysis. *J. Adv. Res.* **2015,** *6* (3), 493–499.
40. Sharma, J.; Balakrishnan, L.; Kaushik, S.; Kashyap, M. K. Multi-Omics Approaches to Study Signaling Pathways. *Front. Bioengg. Biotechnol.* **2020,** *8*, 829.
41. Blum, B. C.; Mousavi, F.; Emili, A. Single-Platform 'Multi-Omic' Profiling: Unified Mass Spectrometry and Computational Workflows for Integrative Proteomics–Metabolomics Analysis. *Mol. Omics* **2018,** *14* (5), 307–319.
42. Bheemanahalli, R.; Knight, M.; Quinones, C.; Doherty, C. J.; Jagadish, S. V.; Genome-Wide Association Study and Gene Network Analyses Reveal Potential Candidate Genes for High Night Temperature Tolerance in Rice. *Sci. Rep.* **2021,** *11* (1), 1–17.
43. Kumar, A.; Thomas, J.; Gill, N.; Dwiningsih, Y.; Ruiz, C.; Famoso, A.; Pereira, A. Molecular Mapping and Characterization of QTLs for Grain Quality Traits in a RIL Population of US Rice Under High Night-time Temperature Stress, 2022.
44. Zhou, H. W.; Zhang, H. Y.; Wang, Z. H.; He, C.; You, L. L.; Fu, D. H.; Song, J. B.; Huang, Y. J.; Liao, J. L.; Discovery of Unique Single Nucleotide Polymorphisms in Rice in Response to High Night-time Temperature Stress Using a Hybrid Sequencing Strategy. *Environ. Exp. Bot.* **2019,** *162*, 48–57.
45. Xu, J.; Misra, G.; Sreenivasulu, N.; Henry, A. What Happens at Night? Physiological Mechanisms Related to Maintaining Grain Yield Under High Night Temperature in Rice. *Plant Cell Environ.* **2021,** *44* (7), 2245–2261.
46. Zhang, T.; Li, X.; Zhao, Z.; Wu, R.; Yang, Z.; He, G. Sequencing and Genomic Analysis of Sorghum DNA Introgression Variant Line R21 and Recipient Rice Jin Hui 1 Revealed Repetitive Element Variation. *Int. J. Mol. Sci.* **2022,** *23* (19), 11864.
47. Chen, L.; Wang, Q.; Tang, M.; Zhang, X.; Pan, Y.; Yang, X.; Gao, G.; Lv, R.; Tao, W.; Jiang, L.; Liang, T. QTL Mapping and Identification of Candidate Genes for Heat Tolerance at the Flowering Stage in Rice. *Front. Genet.* **2021,** *1840.*
48. Lowe, R.; Shirley, N.; Bleackley, M.; Dolan, S.; Shafee, T. Transcriptomics Technologies. *PLoS Computational Biol.* **2017,** *13* (5), 1005457.
49. Dong, Z.; Chen, Y. Transcriptomics: Advances and Approaches. *Sci. China Life Sci.* **2013**. *56* (10), 960–967.
50. Mironova, V. V.; Weinholdt, C.; Grosse, I. RNA-seq Data Analysis for Studying Abiotic Stress in Horticultural Plants. In *Abiotic Stress Biology in Horticultural Plants*; Kanayama, Y., Kochetov, A., Eds.; Springer: Tokyo, 2015; pp 197–220. DOI: 10.1007/978-4-431-55251-2_14

51. Desai, K.; Johnson, J. Virtex: Learning Visual Representations from Textual Annotations. In *Proceedings of the IEEE/CVF Conference on Computer Vision and Pattern Recognition*, 2021; pp 11162–11173.
52. Nusinow, D. A.; Helfer, A.; Hamilton, E. E.; King, J. J.; Imaizumi, T.; Schultz, T. F.; Kay, S. A. The ELF4–ELF3–LUX Complex Links the Circadian Clock to Diurnal Control of Hypocotyl Growth. *Nature* **2011,** *475* (7356), 398–402.
53. W. M. Gray. A.; Ostin, G.; Sandberg, C. P.; Romano, M. Estelle, High Temperature Promotes Auxin-Mediated Hypocotyl Elongation in Arabidopsis. *Proc. Natl. Acad. Sci. USA* **1998,** *95*, 7197–7202
54. Liu, Y. J. Dendritic Cell Subsets and Lineages, and Their Functions in Innate and Adaptive Immunity. *Cell* **2001,** *106* (3), 259–262.
55. Crocco, C. D.; Locascio, A.; Escudero, C. M.; Alabadí, D.; Blázquez, M. A.; Botto, J. F. The Transcriptional Regulator BBX24 Impairs DELLA Activity to Promote Shade Avoidance in *Arabidopsis thaliana. Nat. Commun.* **2015,** *6*, 6202.
56. Lyu, G.; Li, D.; Li, S. Bioinformatics Analysis of BBX Family Genes and Its Response to UV-B in *Arabidopsis thaliana. Plant Signal. Behav.* **2020,** *15* (9), 1782647.
57. Doyal, L. Putting Gender into Health and Globalisation Debates: New Perspectives and Old Challenges. *Third World Quart.* **2002,** *23* (2), 233–250.
58. Guo, J.; Huang, Z.; Sun, J.; Cui, X.; Liu, Y. Research Progress and Future Development Trends in Medicinal Plant Transcriptomics. *Front. Plant Sci.* **2002,** *12*.
59. Arbona, V.; Manzi, M.; de Ollas, C.; Gómez-Cadenas, A. Metabolomics as a Tool to Investigate Abiotic Stress Tolerance in Plants. *Int. J. Mol. Sci.* **2013,** *14* (3), 4885–4911.
60. Xu, S.; Li, J.; Zhang, X.; Wei, H.; Cui, L. Effects of Heat Acclimation Pretreatment on Changes of Membrane Lipid Peroxidation, Antioxidant Metabolites, and Ultrastructure of Chloroplasts in Two Cool-Season Turfgrass Species Under Heat Stress. *Environ. Exp. Bot.* **2006,** *56* (3), 274–285.
61. Yan, S.; Bhawal, R.; Yin, Z.; Thannhauser, T.W.; Zhang, S. Recent Advances in Proteomics and Metabolomics in Plants. *Mol. Hortic.* **2022,** *2* (1), 1–38.
62. Mohammed, A.R.; Tarpley, L. Effects of High Night Temperature and Spikelet Position on Yield-Related Parameters of Rice (*Oryza sativa* L.) Plants. *Europ. J. Agron.* **2010,** *33* (2), 117–123.
63. Glaubitz, U.; Li, X.; Schaedel, S.; Erban, A.; Sulpice, R.; Kopka, J.; Hincha, D.K.; Zuther, E. Integrated Analysis of Rice Transcriptomic and Metabolomic Responses to Elevated Night Temperatures Identifies Sensitivity-and Tolerance-Related Profiles. *Plant Cell Environ.* **2017,** *40* (1), 121–137.
64. Schaarschmidt, S.; Lawas, L.M.F.; Glaubitz, U.; Li, X.; Erban, A.; Kopka, J.; Jagadish, S.K.; Hincha, D.K.; Zuther, E. Season Affects Yield and Metabolic Profiles of Rice (*Oryza sativa*) Under High Night Temperature Stress in the Field. *Int. J. Mol. Sci.* **2020,** *21* (9), 3187.
65. Escandón, M.; Meijón, M.; Valledor, L.; Pascual, J.; Pinto, G.; Cañal, M.J. Metabolome Integrated Analysis of High-Temperature Response in *Pinus radiata. Front. Plant Sci.* **2018,** *9*, 485.
66. Wada, H.; Hatakeyama, Y.; Nakashima, T.; Nonami, H.; Erra-Balsells, R.; Hakata, M.; Nakata, K.; Hiraoka, K.; Onda, Y.; Nakano, H. On-Site Single Pollen Metabolomics

Reveals Varietal Differences in Phosphatidylinositol Synthesis Under Heat Stress Conditions in Rice. *Scient. Rep.* **2020,** *10* (1), 1–11.
67. Xu, Y.; Freund, D. M.; Hegeman, A.D.; Cohen, J.D. Metabolic Signatures of *Arabidopsis thaliana* Abiotic Stress Responses Elucidate Patterns in Stress Priming, Acclimation, and Recovery. *Stress Biol.* **2022,** 1–16.
68. Li, M.; Li, J.; Zhang, R.; Lin, Y.; Xiong, A.; Tan, G.; Luo, Y.; Zhang, Y.; Chen, Q.; Wang, Y.; Zhang, Y. Combined Analysis of the Metabolome and Transcriptome to Explore Heat Stress Responses and Adaptation Mechanisms in Celery (*Apium graveolens* L.). *Int. J. Mol. Sci.* **2022,** *23* (6), 3367.
69. Boehlein, S. K.; Liu, P.; Webster, A.; Ribeiro, C.; Suzuki, M.; Wu, S.; Guan, J. C.; Stewart, J. D.; Tracy, W. F.; Settles, A. M.; McCarty, D. R. Effects of Long-Term Exposure to Elevated Temperature on Zea Mays Endosperm Development During Grain Fill. *Plant J.* **2019,** *99* (1), 23–40.
70. Zhao, C.; Zhang, Y.; Du, J.; Guo, X.; Wen, W.; Gu, S.; Wang, J.; Fan, J. Crop Phenomics: Current Status and Perspectives. *Front. Plant Sci.* **2019,** *10*, 714.
71. Yang, W.; Feng, H.; Zhang, X.; Zhang, J.; Doonan, J. H.; Batchelor, W. D.; Xiong, L.; Yan, J. Crop Phenomics and High-Throughput Phenotyping: Past Decades, Current Challenges, and Future Perspectives. *Mol. Plant.* **2020,** *13* (2), 187–214.
72. Deery, D. M.; Jones, H. G. Field Phenomics: Will It Enable Crop Improvement? *Plant Phenom.* **2021.**
73. Lekshmy, V. S.; Vijayaraghavareddy, P.; Nagashree, A. N.; Ramu, V. S.; Ramegowda, V.; Makarla, U.; Sreeman, S. Induction of Acquired Tolerance Through Gradual Progression of Drought Is the Key for Maintenance of Spikelet Fertility and Yield in Rice Under Semi-Irrigated Aerobic Conditions. *Front. Plant Sci.* **2021,** *2275.*
74. Chen, S.; Guo, Y.; Sirault, X.; Stefanova, K.; Saradadevi, R.; Turner, N. C.; Nelson, M. N.; Furbank, R. T.; Siddique, K. H.; Cowling, W. A. Non Destructive Phenomic Tools for the Prediction of Heat and Drought Tolerance at Anthesis in Brassica Species. *Plant Phenom.* **2019.**
75. Hein, A.; Weking, J.; Schreieck, M.; Wiesche, M.; Böhm, M.; Krcmar, H. Value Co-Creation Practices in Business-to-Business Platform Ecosystems. *Electron. Markets* **2019,** *29* (3), 503–518.
76. Hein, A.; Schreieck, M.; Riasanow, T.; Setzke, D. S.; Wiesche, M.; Böhm, M.; Krcmar, H. Digital Platform Ecosystems. *Electron. Markets* **2020,** *30* (1), 87–98.
77. Basavaraj, P.S.; Rane, J. Avenues to Realize Potential of Phenomics to Accelerate Crop Breeding for Heat Tolerance. *Plant Physiol. Reports.* **2020,** *25* (4), 594–610.
78. Sreeman, S.M.; Vijayaraghavareddy, P.; Sreevathsa, R.; Rajendrareddy, S.; Arakesh, S.; Bharti, P.; Dharmappa, P.; Soolanayakanahally, R. Introgression of Physiological Traits for a Comprehensive Improvement of Drought Adaptation in Crop Plants. *Front. Chem.* **2018,** *6*, 92.
79. Mori, K.; Sugaya, S.; Gemma, H. Decreased Anthocyanin Biosynthesis in Grape Berries Grown Under Elevated Night Temperature Condition. *Sci. Hortic.* **2005,** *105* (3), 319–330.
80. Soltani, A.; Weraduwage, S. M.; Sharkey, T. D.; Lowry, D. B. Elevated Temperatures Cause Loss of Seed Set in Common Bean (*Phaseolus vulgaris* L.) Potentially Through the Disruption of Source-Sink Relationships. *BMC Genom.* **2019,** *20* (1), 1–18.

Index

M

N

O

Y